Economic Liberalization, Democratization and Civil Society in the Developing World

International Political Economy Series

General Editor: **Timothy M. Shaw**, Professor of Political Science and International Development Studies, and Director of the Centre for Foreign Policy Studies, Dalhousie University, Halifax, Nova Scotia

Titles include:

Francis Adams, Satya Gupta and Kidane Mengisteab (*editors*)
GLOBALIZATION AND THE DILEMMAS OF THE STATE IN THE SOUTH

Susan Dicklitch
THE ELUSIVE PROMISE OF NGOs IN AFRICA
Lessons from Uganda

David Hulme and Michael Edwards (*editors*)
NGOs, STATES AND DONORS
Too Close for Comfort?

Staffan Lindberg and Árni Sverrisson (*editors*)
SOCIAL MOVEMENTS IN DEVELOPMENT
The Challenge of Globalization and Democratization

Laura MacDonald
SUPPORTING CIVIL SOCIETY
The Political Role of Non-Governmental Organizations in Central America

Kurt Mills
HUMAN RIGHTS IN THE EMERGING GLOBAL ORDER
A New Sovereignty

Michael G. Schechter (*editor*)
THE REVIVAL OF CIVIL SOCIETY
Global and Comparative Perspectives

International Political Economy Series
Series Standing Order ISBN 0–333–71708–2 hardcover
Series Standing Order ISBN 0–333–71110–6 paperback
(*outside North America only*)

You can receive future titles in this series as they are published by placing a standing order. Please contact your bookseller or, in case of difficulty, write to us at the address below with your name and address, the title of the series and one of the ISBNs quoted above.

Customer Services Department, Macmillan Distribution Ltd, Houndmills, Basingstoke, Hampshire RG21 6XS, England

Economic Liberalization, Democratization and Civil Society in the Developing World

Edited by

Remonda Bensabat Kleinberg
Assistant Professor
Department of Political Science
University of North Carolina at Wilmington
USA

and

Janine A. Clark
Assistant Professor
Department of Political Science
University of Guelph
Ontario
Canada

palgrave

Published by PALGRAVE
Houndmills, Basingstoke, Hampshire RG21 6XS and
175 Fifth Avenue, New York, N. Y. 10010
Companies and representatives throughout the world

PALGRAVE is the new global academic imprint of
St. Martin's Press LLC Scholarly and Reference Division and
Palgrave Publishers Ltd (formerly Macmillan Press Ltd).

Outside North America
ISBN 0–333–72068–7

In North America
ISBN 0–312–23087–7

This book is printed on paper suitable for recycling and made from fully managed and sustained forest sources.

A catalogue record for this book is available from the British Library.

Library of Congress Catalog Card Number: 99–053108

Transferred to digital printing 2002

Printed and bound in Great Britain by
Antony Rowe Ltd, Chippenham, Wiltshire

Contents

Acknowledgments

There are many people and institutions to whom we are indebted in the process of researching, writing and editing this book. First among them are the Social Science and Humanities Research Council of Canada, and the Center for the Humanities, University of New Hampshire, for enabling us to conduct our work at every stage of the process. Funds from the Summer Initiative program from the University of North Carolina at Wilmington were also extremely helpful during the research stage for sections of this book. Special thanks must go to Tim Shaw, the editor of this series in Political Economy. Rex Brynen, Richard Sandbrook, Janice Stein, Denis Sullivan, and Homa Hoodfar must also be thanked for the advice and support they regularly provided us.

Both Joel Migdal from the University of Washington, and Jonathan Hartlyn from the University of North Carolina, Chapel Hill, were invaluable as Chair and Discussant respectively, at our preliminary panel at the annual American Political Science Association conference in San Francisco. Both distinguished scholars offered constructive advice and were warmly accessible. Without Allison Howard this book could not have been completed. Her hard work, diligence and conscientiousness at seemingly endless proof-reading was invaluable. We owe Dee-Ann Dickson for her laughter and her last-minute typing. Dave Pelletier must be credited for compiling our index. We are also indebted to Funso Afolayan and Stephen J. King for helping us out at the last minute. Finally, Howard Kleinberg must be commended for his patience, helpful ear, and great cooking skills.

R. B. K.
J. A. C.

List of Abbreviations

ABIM	Islamic Youth Movement Malaysia
ABM	Mexican Bankers' Association
ACC	Amman Chamber of Commerce
ACFODE	Action for Development (Uganda)
ACI	Amman Chamber of Industry
AD	Alliance for Democracy (Nigeria)
ADGGN	Good Governance in Nigeria
ALG	Alternative Law Groups (Philippines)
AMNLAE	Nicaraguan Association of Women Luisa Amanda Espinoza
APEC	Asia-Pacific Economic Cooperation
APP	All People's Party (Nigeria)
ARE	Asian Rare Earth Sdn Bhd (Malaysia)
ASN	Permodalan Amanah Saham Association (Malaysia)
BAYAN	New Patriot Alliance (Philippines)
BMF	Bumiputera Malaysia Finance
BN	Barisan Nasional (National Front, Malaysia)
CANACINTRA	Chamber of Transformation Industries (Mexico)
CAP	Consumers Association of Penang (Malaysia)
CGT	General Confederation of Labor (Argentina)
CHR	Commission on Human Rights (Philippines)
CIDA	Canadian International Development Agency
CLO	Civil Liberty Organization (Algeria)
CLOAs	Certificates of Land Ownership Award (Philippines)
CM	Council Members (Uganda)
CNC	National Peasant Confederation
COECE	Coordinating Council for the Free Trade Agreement (Mexico)
CONAPRO H-M	Women's Secretariat of the Confederation of the Nicaraguan Professionals Heroes and Martyrs
CONCAMIN	Confederation of Chambers of Industry (Mexico)
CONCANACO	Confederation of Chambers of Commerce (Mexico)

CP	Conservative Party (Uganda)
CPK	Church of the Province of Kenya
CPLA	Cordillera People's Liberation Army (Philippines)
CTA	Congress of Argentinean Workers
CTM	Mexican Workers Confederation
DA	District Administrator (Uganda)
DAP	Democratic Action Party (Malaysia)
DNT	National Department of Labour (Argentina)
DP	Democratic Party (Kenya), (Uganda)
ECC	Economic Consultative Council (Jordan)
EPR	People's Revolutionary Army (Mexico)
EPSM	Environmental Protection Society Malaysia
ERP	Economic Recovery Program (Uganda)
ERSAP	Economic Reform and Structural Adjustment Program (Egypt)
ESAF II	Enhanced Structural Adjustment Facility (Nicaragua)
ESRC	Economic and Social Research Council
EZLN	Zapatista Army of National Liberation (Mexico)
FAD	Foundation for African Development (Uganda)
FBSI	All-Indonesian Labor Federation
FIDA	International Association of Women Lawyers (Uganda Branch)
FIMA	Food Industries of Malaysia
FIND	Families of Victims of Involuntary Disappearances (Philippines)
FLAG	Free Legal Assistance Group (Philippines)
FOMCA	Federation of Malaysian Consumers Association
FORD	Forum for the Restoration of Democracy (Kenya)
FSLN	Sandinista Front for National Liberation (Nicaragua)
GABRIELA	General Assembly Binding Women for Reform, Integrity, Equality, Leadership and Action (Philippines)
GATT	General Agreement on Tariffs and Trade
GDP	Gross domestic product
GNP	Gross national product
GSO	Government sponsored organizations
GSP	Generalized System of Preferences
HICOM	Heavy Industries of Malaysia
ICA	Industrial Coordination Act (Malaysia)

ICFTU	International Confederation of Free Trade Unions (Indonesia)
IFI	International financial institutions
IGG	Inspector-General of Government (Uganda)
ILO	International Labor Organization
IMF	International Monetary Fund
INFID	Forum on Indonesian Development
IPFC	Inter-Party Forces Committee (Uganda)
IPK	Islamic Party of Kenya
ISA	Internal Security Act (Malaysia)
ISI	Import Substitution Industrialization
IWK	Indah Water Konsortium (Malaysia)
JIC	Jordan Investment Corporation
KADU	Kenya People's Democratic Union
KANU	Kenya Africa National Union
KMP	Philippine Peasant Movement
KMU	May First Movement (Philippines)
KOPKAMTIB	Command for the Restoration of Order and Security (Indonesia)
KPU	Kenya People's Union
LC	Local Council (Uganda)
LRA	Lord's Resistance Army (Uganda)
MARTYR	Mothers and Relatives against Tyranny and Repression (Philippines)
MAS	Malaysian Airlines System
MCA	Malaysian Chinese Association
MCP	Malawi Congress Party
MFPA	Manila People's Forum on APEC
MHRC	Moro Human Rights Center (Philippines)
MIC	Malaysian Indian Congress
MISC	Malaysian International Shipping Corporation
MMD	Movement for Multi-Party Democracy (Zambia)
MNLF	Moro National Liberation Front (Philippines)
MOS	Ministry of Supplies (Jordan)
MOSA	Ministry of Social Affairs (Egypt)
MTA	Movement of Argentinan Workers
NAFTA	North American Free Trade Agreement
NBI	National Bureau of Investigation (Philippines)
NCCK	National Council of Churches of Kenya
NCPD	National NGO Committee for Population Development (Egypt)

NDF	National Democratic Front (Philippines)
NDP	National Democratic Party (Egypt)
NDP	National Development Policy (Malaysia)
NEP	New Economic Policy (Malaysia)
NGO	Non-governmental organization
NIJU	National Institute of Journalists of Uganda
NOCEM	National Organization for Civic Education and Election Monitoring (Uganda)
NOTU	National Organization of Trade Unions (Uganda)
NPA	New People's Army (Philippines)
NRA	National Resistance Army (Uganda)
NRC	National Resistance Council (Uganda)
NRC	National Republican Convention (Nigeria)
NRM	National Resistance Movement (Uganda)
PAHRA	Philippine Alliance of Human Rights Advocates
PARC	Perak Anti-Radioactivity Committee
PAS	Islamic Party of Malaysia
PBDS	Parti Bansa Dayak Sarawak (Malaysia)
PBS	Parti Bersatu Sabah (Malaysia)
PDI	Indonesian Democratic Party
PHRC	Presidential Human Rights Commission (Philippines)
PKI	Indonesian Communist Party
PLLP	Protestant Lawyers League of the Philippines
PNB	National Equity Corporation (Malaysia)
PNR	National Revolutionary Party (Mexico)
POs	People's organizations
PPBI	Center for Indonesian Workers' Struggle
PRD	People's Democratic Alliance (Indonesia)
PRD	Party of the Democratic Revolution (Mexico)
PRI	Institutional Revolutionary Party (Mexico)
PRONASOL	Program for National Solidarity (Mexico)
PU	Privatization Unit (Uganda)
RCD	Democratic Constitutional Rally (Tunisia
RIC	Regional Industrial Centers (Philippines)
SAFINA	The Ark
SAL	Structural Adjustment Loans
SALAG	Structural Alternative Legal Assistance for Grassroots (Philippines)
SAP	Structural Adjustment Program

SAPRIN	Structural Adjustment Participatory Review Network (Uganda)
SBSI	Prosperous Indonesia Workers' Union
SEDCs	State Economic Development Corporations (Malaysia)
SELDA	Association of Ex-Detainees against Detention and for Amnesty (Philippines)
SGS	Selangor Graduates Society (Malaysia)
TCC	Telecommunications Corporation (Jordan)
TFDP	Task Force Detainees of the Philippines
UDA	Urban Development Authority (Malaysia)
UDF	United Democratic Front (Malawi)
UGIETA	Uganda Importers and Exporters Association
UHRA	Uganda Human Rights Activist
UJA	Ugandan Journalists Association
UMNO	United Malays National Organization
UNLA	Uganda National Liberation Army
UNO	National United Opposition (Nicaragua)
UPC	Uganda People's Congress
UPM	Uganda Patriotic Movement
USAID	United States Agency for International Development
UYD	Uganda Young Democrats
VAT	Value added tax
WALHI	Indonesian Environmental Forum
WNBF	West Nile Bank Front (Uganda)
WTO	World Trade Organization

Notes on the Contributors

Funso Afolayan holds a PhD in African history from Obafemi Awolowo University, Ile Ife, Nigeria. He is co-author of *Yoruba Sacred Kingship: A Power Like That of the Gods*, and has contributed to numerous edited books. He currently teaches African and world history at the University of New Hampshire, Durham.

Remonda Bensabat Kleinberg, PhD, University of Toronto, Canada, is an Assistant Professor at the University of North Carolina at Wilmington. In addition to numerous articles she is the author of *Strategic Alliances and Other Deals: Business–State Relations and Economic Reform in Mexico.*

Janine A. Clark, PhD, University of Toronto, Canada, is an Assistant Professor at the University of Guelph where she teaches Middle Eastern politics. Currently she is conducting research on women in Islamist political parties in Yemen and Jordan.

Gerard Clarke is a Lecturer in Development Studies at the University of Wales, Swansea. He is the author of *The Politics of NGOs in South-East Asia: Participation and Protest in the Philippines*. He has undertaken field research on NGOs in the Philippines, Thailand, Cambodia, Vietnam and Ethiopia.

Susan Dicklitch, PhD, University of Toronto, Canada, is author of *The Elusive Promise of NGOs in Africa: Lessons from Uganda*. She teaches African politics at Franklin Marshall University, Pennsylvania. Her present research focuses on Ghana.

Katherine Isbester received her PhD from the University of Toronto in political science. She has published numerous articles on the women's movement in Nicaragua. She has lectured at the University of Toronto and now resides in London, England.

Sidney Jones is the Executive Director of the Asia Division of Human Rights Watch and is on the board of the National Committee of US–China Relations. She has worked as a researcher and program officer at Amnesty International and the Ford Foundation, Jakarta, respectively. Ms Jones has written extensively on human rights in Asia with a particular focus on Indonesia.

Stephen J. King is Assistant Professor of Government at Georgetown University. He is currently working on a manuscript on the politics of market reforms in Tunisia.

Francis Kok-Wah Loh (BA, Dartmouth; PhD, Cornell) is an Associate Professor of Political Science at the Universiti Sains Malaysia, Penang. He has published widely on Malaysian and Southeast Asian affairs. His current research focuses on the politics of developmentalism and the middle classes in Southeast Asia.

Pete W. Moore, PhD, Department of Political Science, McGill University, Montreal, has received research awards from the Social Science Research Council, the Fulbright Foundation, and the American Centers of Oriental Research. He is currently working on a project concerning competitive rent-seeking among business groups in several Arab states.

Julius E. Nyang'oro is a Professor and Chair of African and Afro-American Studies at the University of North Carolina at Chapel Hill. He has published widely in the field of African political economy, and regularly works with NGOs in eastern and southern Africa. His main interest at the moment is on African governance.

Viviana Patroni teaches political science at Wilfrid Laurier University, Ontario, Canada. She received her PhD from the Department of Social and Political Thought, York University, Toronto. Her current research foci are on state–labor relations in Mexico and Argentina, and on the political economy of development in Latin America.

Introduction: The Impact of Economic Reform on Civil Society, Popular Participation and Democratization in the Developing World

Janine A. Clark and Remonda Bensabat Kleinberg

Civil society: popular conceptions and definitions

The term "civil society" has experienced a resurgence in policy making and scholarly circles in the past ten years. International financial institutions are no exception. Terms such as empowerment, participation and democratization are often coined in conjunction with civil society – connoting the positive and progressive role of civil society in the developing world. Indeed, there has been a marked shift from the view of civil society – those autonomous organizations that lie between society and the state – as playing a positive and necessary role in the alleviation of poverty to one in which society is also seen as essential to popular participation and democratization.

Both the revival of this term and its new-found usage are not without their critics. Rejectionists of the term point to its diverse intellectual origins, and its differing definitions and usages. The roots of the debate over the meaning and role of civil society are found in the two different intellectual origins of the term. The first is generally labeled as liberal, Lockian or Anglo-American, and the second as post-Hegelian or Marxist.[1] Classical liberals deem civil society as natural or inherently good. It is a sphere of autonomy and liberty; a counter to state power. Individuals are viewed equal in their possession of rights, and their ability to use them to ensure their personal freedoms.[2] Marxists understand it as "either a step on the way to state formation or a handmaiden in the state's project of domination."[3] Moreover, as the Marxist view emphasizes, modern civil societies that provide rights to each person "do not encourage individuals to use these rights to shape the political community and to forge the direction of the economy. The major economic decisions are left to an elite few, those who are the owners or managers of capital."[4] It is therefore integral to class oppression.

Furthermore, political philosophers, such as Alexis de Tocqueville, John Locke, Karl Marx, and Antonio Gramsci, amongst others, essentially representing these two traditions also differ amongst themselves. Little agreement can be found, for example, on whether civil society is a "public space," associations or norms, on the so-called division between civil society and the state, on the nature of civil society (in other words, its "civility"), or on its relationship to economic change and/or the class struggle.

The World Bank, the International Monetary Fund (IMF), and other international financial and aid institutions (IFIs) explicitly treat civil society as synonymous with private-sector associations, including business organizations. Civil society is treated as equal and just or "civil" and a clear distinction is understood between civil society and the state.

Structural adjustment and the "politics" of civil society: assumptions of democratization

The World Bank furthermore locates civil society within a larger perspective that regards economic liberalization, political liberalization, and democratization as mutually reinforcing processes. Economic restructuring programs are thus deemed as compatible, indeed integral, to the introduction of political liberties and, ultimately, democracy.

Alternatively, the assumption is that governments in the developing world are significantly legitimized through an electoral process, and their primary role lies in merely withdrawing from the economic sphere as they implement economic liberalization measures:

> Governments of *different persuasions* are looking for ways to freeze and eventually reduce [this] spending *without losing votes* [emphasis ours]. The market is often the most acceptable and efficient solution – notwithstanding the severe Asian economic crisis of 1998 – with private health and unemployment insurance, private contributions to higher education costs and the care of the elderly and even private prison services among reforms initiated in the 1990s.[5]

The unproblematic association between a strong civil society, export-oriented capitalism and democracy can be found as the premise upon which international aid agencies base their programs. In its preamble the United States Agency for International Development (USAID) states that "sustainable development results from: the implementation of open, market-oriented economics and institutions; social policies that increase human capacity and opportunities for individuals to improve their lives; open and accessible political institutions and processes that encourage the active engagement of all members of society."[6] USAID suggests that "each of these conditions contributes to the success of all the others," and the lack of one or more of these conditions threatens the success of all the others.[7]

In its own missions the Canadian International Development Agency (CIDA), for example, clearly reflects this positive correlation:

> While the state provides the various forms and organisations of government to carry out public functions, it is civil society which creates the values and normative framework for governance and private enterprise. Its function is to create the social conscience and accepted social consensus within which the formal government structure and private enterprise is to operate. Without a well-rooted civil society, private enterprise, human rights, democratic development and good governance cannot flourish.[8]

Broadly speaking, civil society's role in this understanding of the transition to democracy is three-fold. As the state withdraws economically, civil associations aid in meeting the needs of the poor and reducing the demands on the state. In terms of sheer numbers, civil society comprises increasing voices pressuring the state for a decentralization of decision-making powers. Finally, civil society is deemed to play a critical role in the creation of a participatory civic culture. Civic associations help to create a civic culture through their leadership or demonstrative role in organizing and demanding policy changes. They also do so by projecting an alternative vision and organization of society. Hence the democratic and participatory manner in which decisions are made in civic associations, and their coterminous organizational structures, help promote a democratic civic culture.

In sum, associations are deemed to foster participatory experiences, leadership skills and democratic values. They thereby prepare people for democratic political activity and press them to demand democratic change in all aspects of their lives. In this manner the democratic ideal is deemed to spread, take root and ultimately bring about political change.

The link between structural adjustment and democratization as visualized by the IMF, the World Bank and other international financial

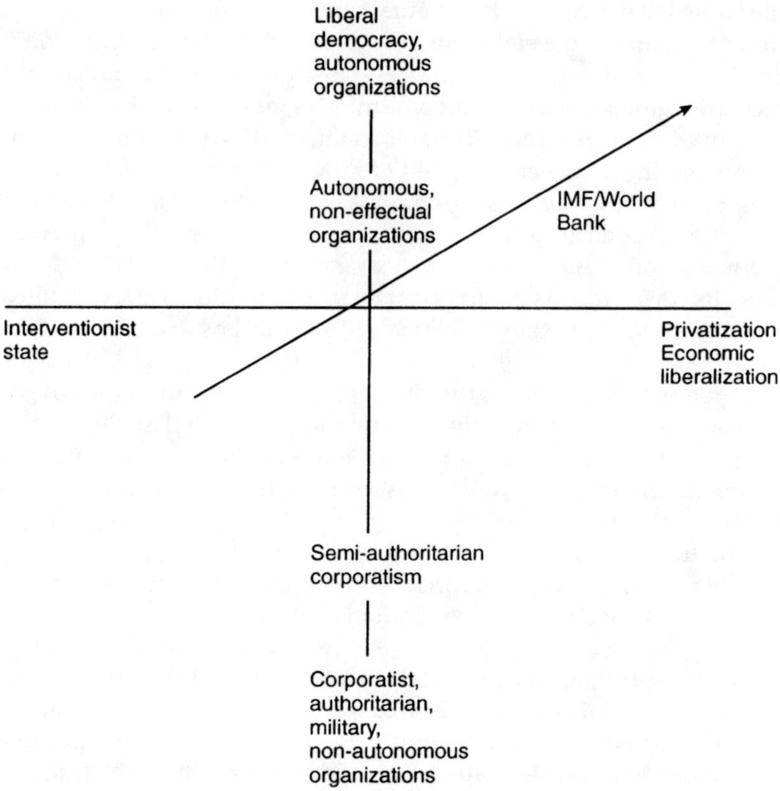

Fig. 1 IFI projections, 1980–99

and aid agencies can be plotted simply as Figure 1 suggests. Here, as privatization accelerates, so does the advent of liberal democracy.

Challenges to those assumptions

Increasingly, scholars are challenging this assumed positive causal relationship between economic reform and democratization. In his article, "Crises, Elites, and Democratization in the Arab World," Saad Eddin Ibrahim argues that, despite the economic and political crises that have afflicted authoritarian regimes of the Middle East, they have been able to avoid democratization via four possible "survival strategies." Regimes can use coercion to repress opposition to state policies, although this option has been "stretched to the limits of diminishing returns." They can choose to contrive or take advantage of a real crisis to deflect the opposition towards a different target; this too grants only limited breathing space. The third option centers on symbolic or token power sharing with select opposition groups and the fourth on symbolic participatory politics. While Arab authoritarianism is eroding, democratization is not necessarily proceeding.[9]

Speaking directly to the Somali crisis in Africa, Julius O. Ihonvbere suggests that structural adjustment programes imposed by IMF/World Bank regulations actually *exacerbated* non-democratic and violent conditions. He argues that the "IMF and the World Bank waded in in the 1980s only to deepen contradictions, destroy the foundations of stability, erode the legitimacy of the state, intensify poverty and alienation and lay the foundation for the more popularly known version of the Somali experience as was seen very recently."[10]

Even post-authoritarian Latin America is no exception to the rule, despite its success rate in the mid-1990s. Reviewing the links between structural adjustment, markets, and democracy, editors William Smith, Carlos Acuña and Eduardo Gamarra observe that:

> [t]he postauthoritarian regimes that emerged in Latin America over the past decade have achieved a degree of democratic consolidation, but they continue to confront severe political and economic challenges. Stable and sustainable growth remains an elusive goal for many countries, despite the modest successes of neoliberal strategies of market-oriented restructuring in achieving external adjustment, taming hyperinflation, and disciplining public sector finance. Sharp declines in per capita income in many countries have exacerbated poverty and worsened already egregious patterns of inequality and concentration of wealth and control over productive resources. Moreover, the often autocratic implementation frequently associated with stabilization and structural adjustment policies has shown a disturbing tendency to undermine representative institutions and the conditions for deepening democratic politics, particularly in the broadening of citizenship rights beyond the electoral arena.[11]

The book

The authors in this edited volume further challenge the assumption that economic reform will necessarily lead to greater democratization. In particular, we question whether economic structural adjustment will translate into the withdrawal of the state at the grassroots level, and whether, as a result, political "space" will be created for greater numbers of non-state actors to express their voices and more directly influence policy-making. The book is therefore an exploration of the political impact of economic structural adjustment. The contributors examine the impact of macroeconomic policies upon micropolitical processes and, in turn, the significance of these microchanges for macropolitical changes towards greater democratization. In other words, what do changes, if any, at the micro level – the space called civil society – mean for the processes of political liberalization and democratization at the macro level?

To do so we have brought together twelve countries from four regions of the developing world, Asia, Africa, the Middle East and North Africa, and Latin America, and take an explicitly grassroots look at their political transitions within the context of economic restructuring. In the first two chapters of the volume, Sidney Jones and Gerard Clarke examine human rights non-governmental organizations (NGOs) in Indonesia and the Philippines respectively.

In the first chapter (on Indonesia), Sidney Jones observes that, up until May 1998, forces for democratization outside the NGO community, and to some extent within it, remained weak. As Jones stresses, while the middle class resented the cronyism and corruption of the Soeharto regime (1969–98), it provided little concrete resistance. The advent of rapid economic growth (1971–82) and the corresponding concentration of economic and political power in the hands of the Soeharto family, however, led to the forging of "some of the best known NGOs in Indonesia," the number and visibility of which increased dramatically during the 1980s. This period also coincided with the beginning of Indonesia's structural adjustment process. Jones argues that while the structural adjustment program (SAP) was not totally responsible for the growth of NGOs *per se*, the impact of restructuring directly and indirectly influenced NGO advocacy in the area of human rights, the environment, and labor rights. The latter particularly became one of the major issues for NGOs in the 1990s but, until recently, the combination of state repression and state concessions limited their ability to organize openly in factories.

In Chapter 2, Gerard Clarke traces the changes in the nature of human rights advocacy, the human rights NGO community, and the relationship between human rights NGOs and the state as economic and political reforms have progressed in the Philippines. He argues that while human rights NGOs played an important role in the introduction of democracy in the Philippines, as electoral democracy has consolidated they have been increasingly marginalized. As long as human rights NGOs confined their activities to abuses of individual freedoms they were able to wield considerable influence and leverage *vis-à-vis* the state. With their re-definition of human rights to include "development aggression" – human rights violations that are not political in nature but inflicted as part of the inequitable model of development and economic growth – their ability to influence policy has been eroded.

In Chapter 3, Francis Kok-Wah Loh focuses on business groups in Malaysia. Loh examines the rise of a multi-ethnic middle class as a result of Malaysia's New Economic Policy and its political impact upon democratization. Throughout the 1980s the new middle classes provided the motor for NGO activity and were actively engaged in pressuring the state for greater political liberalization and democratization. In the 1990s, however, he argues that this has come to a halt as a result of three interrelated processes. The first is the limits to state "flexibility" and the imposition of bans on civil society activity. The second is the state's ample use of patronage to coopt certain sectors, particularly big business. The third is the development of a culture of "developmentalist" – one which valorizes consumerist habits dependent upon rapid growth – as a result of economic reforms. The government also directed economic policies toward the more "underprivileged" sectors of society allowing for a generalized acceptance of structural adjustment programs. Loh argues that the culture of mass consumerism that has developed as a result of economic reform has played a role in weakening civil society's effectiveness. Malay society, thus, has largely acquiesced to the consolidation of executive power that has taken place particularly since 1987 when demands for participatory democracy, accountability and transparency proved to be too great for the state to accept.

An examination of political parties in Kenya as the main focal point for civil society opens the section of Africa. Julius E. Nyang'oro looks at the role civil society has played in the transition to democracy in Kenya – both at the formal and informal levels. Nyang'oro argues that, while civil society has managed to play an important role in political liberalization, ethnicity continues to be a divisive force, used for narrow political ends. Informal groups are still in their infancy and remain weak and largely fractured. Indeed, civil society has become "a nasty place which reflects contradictions in all society," according to him.

Within this context of violence, the church and student groups have managed to become particularly active and influential in Kenya.

Despite this, however, the burden of pressuring for democratization has fallen to the newly created political parties. The lack of non-violent channels and non-ethnically based groups has meant that political parties have used ethnicity as a tool of mobilization. Political parties, as a result, have not yet been able to mount a united and effective opposition front to the ruling party. The lack of political parties based on ideology – as opposed to ethnicity – has hindered the progress of democratization.

Violence and ethnicity have plagued much of Uganda's emergence as a state and, like Kenya, internal weaknesses and external controls plague much of Ugandan civil society resulting in economic liberalization without political liberalization. In the second African case study Susan Dicklitch suggests that "there are three main problems associated with civil society and democratization in Africa: the historical impediments to the vibrant growth of a strong and democratic civil society, the weakness and lack of democratic rule of many associations within civil society, and regime impediments." While the growth of NGOs since 1986 has been massive, Yoweri Museveni and his National Resistance Movement (NRM) have succeeded in controlling NGOs, political parties, trade unions, the media and several religious organizations, as well as restricting autonomous groups.

The Ugandan regime's "carrot and stick" approach and its ability to co-op individuals and groups have succeeded in undermining any real political activity the NGO umbrella organization – the National Organization for Civic Education and Election Monitoring (NOCEM) – could instigate. NOCEM and its 14 NGO membership, including human rights, environmental, and women's organizations, were essentially labelled as "criminal" by the NRM regime.

Fully embracing structural adjustment, Museveni himself has spoken of the trade-off between economic growth and stability, and the long-term commitment of democratization. Stability, however, was maintained through a top-down structure of Local Councils (LCs) controlled by the NRM. According to Dicklitch a trade-off has effectively been made. Indeed, "Western donors and the IMF and World Bank have not pressured Uganda to adopt greater political liberalization, mainly because Uganda has so successfully and wholeheartedly implemented SAPS."

While Kenya, and to some extent Uganda, continue to be influenced by "ethno-sectarian antipathies," Funso Afolayan rejects this thesis in his examination of Nigeria. As early as 1981, pro-democracy and human rights groups began emerging as a response to increasing authoritarianism and intra-elite struggles. Afolayan's detailed analysis focuses particularly on popular culture as the cornerstone of a renewed civil society, and its expression through the popular press, electronic media, art, and song. Popular culture intensified with the negative economic effects of structural adjustment, the falling naira, rising external debt, and the extreme drop in per-capita income. Rather than allow for a smoother transition to more democratic structures, the effects of economic reform led to "strikes, industrial unrest, graft, corruption, and blatant kleptocracy at every level of governance."

Popular pro-democracy sentiment disseminated through music in all areas of the country made it difficult for the administration to clamp down on the dissenting artists or supporters. But, by 1993, the military did reinstall temporary rule. The arena for political discourse, however, was, according to Afolayan, significantly expanded.

In the first case study from the Middle East, Janine Clark focuses on social-welfare NGOs in Egypt. Clark examines two interrelated aspects of economic and political reforms as they affect social-welfare NGOs. The first is NGOs' service capacity and sustainability – their ability to deal with the increasing numbers of people negatively affected by economic reform and to manage greater social services and health-care responsibilities as the state withdraws from the arena. The second is whether or not the political space is being created within which NGOs can wield greater policy – and decision-making influence. She argues that the majority of NGOs are too small, often highly dependent upon state and foreign aid, and lacking in trained staff and organizational skills to be able to effectively address the needs of increasing numbers of poor. More importantly, the continuation of state restrictions and regulations has meant that greater political space has not been created in the wake of economic withdrawal. To the contrary, as privatization proceeds there is increasing evidence that ministerial attempts to control the sector may intensify.

In his examination of Jordanian business groups, in Chapter 8, Peter Moore argues that economic reform has brought little meaningful dialog between private interests and the government in macro-policy. Business groups' lack of effective political influence – despite a declared program of privatization – is a result of two primary factors. The first concerns the divided nature of business interests according to sectors and the origins of the merchants themselves. Business groups have not been able to form a pro-business political party in parliament or an effective lobby. The second reason lies in the controlled and limited nature of economic and political liberalization by the state. As Moore states, privatization in Jordan is more of a direction than a plan, with few formal avenues through which business can make its voice heard.

In the third Middle East case study, Stephen King focuses on the impact of neoliberal reforms on the elements of civil society that enhance the prospects for democratization in the North African country of Tunisia. King offers a micro-level analysis and looks closely at the rural community of Tebourba, a large village in Tunisia's rich northwest agricultural region. King argues that neoliberal reforms favored large landowners at the expense of small and medium peasants. Civil society in general in Tunisia has become less vibrant during state-led economic liberalization and the forms of civil society thought to be the most helpful for democratization suffer while potentially detrimental forms are promoted.

In the first case study on Latin America, Remonda Bensabat Kleinberg argues that economic liberalization in Mexico has resulted in greater social inequalities and political instability, and not in political liberalization *per se*. Economic reforms, culminating in the signing of the North American Free Trade Agreement, have entailed the abandonment of the state's traditional "social pact" and the creation of a new *Pacto* or *Allianza* that has dramatically increased big business' control over the direction of the economy. The result has been the concentration of benefits and representation to big business at the expense of small and medium business and labor. The semi-authoritarian corporatist structure remains in place with big business renegotiating it to their favor. Unlike business groups in Jordan, big business has successfully used economic reform to its advantage.

In the only chapter in the volume to look specifically at labour groups, Viviana Patroni studies the case of Argentina. Patroni examines the sophisticated way in which the Peronist government – traditionally the champion of labor – has been able to impose unpopular economic reforms. Precisely because the government is Peronist it has had greatest success in implementing changes without massive labor unrest. While maintaining the traditional corporatist structure, and thus ensuring both labor's access to the state and the state's control over labor, and introducing strict new labor laws, the government has been able to neutralize (but not alienate) labor and avoid a destabilization of the government. As part of the same process, labor has been divided and weakened. The result, according to Patroni, is that the democratization of neither the state nor labor has occurred.

Finally, in Chapter 12, Katherine Isbester looks at women's groups in Nicaragua. Isbester takes an in-depth grassroots look at the women's movement comparing it under both the Sandinista regime (1979–90) and the Violeta Chamorro Administration (1990–6). In doing so, Isbester examines both its internal structure and its relationship to the government/state. Three important trends emerge from her analysis. The first is the declining social and economic conditions for Nicaragua's women since the implementation of market economic reforms in post-socialist Nicaragua. The second is the changing structure of the women's movement itself. In reaction against the controlled and hierarchical nature of the movement under the Sandinistas, the women's movement has taken advantage of the political space offered to it under the new regime and has created a horizontal, democratic internal structure. Its decision-making structure and, technically, its relationship to the state are thus more democratic and autonomous. These positive developments, however, are offset by the fact that greater autonomy has not been matched by greater effectiveness. In fact, the women's movement has lost all of its avenues to state policy-making.

The dominant conservative values ushered in by the Chamorro regime provided subtle yet powerful restraints to the influence of women activists. Women thus confront a new form of intervention in their private lives through new legislation banning rights to abortions, for example, and a government that is hostile to their movement. Isbester emphasizes how the implementation of structural adjustment programs (SAPS) and Catholic policies under the Chamorro regime combined to significantly weaken the more autonomous women's movement.

Contrary to conventional wisdom, our collective analyses substantiate and expand upon critics' claims that political withdrawal rarely accompanies economic withdrawal. Rather than the assumed democratization, we often find increasing authoritarianism. Neither a devolution of state corporatist structures nor a required decrease in coercion necessarily accompanies structural adjustment. In fact, the authors in this volume highlight a fundamental tension between economic liberalization and democratization. Continued and/or increased authoritarianism appears to be the preferred option on behalf of governments in order to force painful economic changes and silence opposition voices.

The most powerful example is provided by the Argentinian case where the Peronist party in power – a party historically based on a platform of workers' rights – has introduced drastic anti-labor legislation in order to ensure the successful implementation of painful economic reforms. As Patroni indicates in her analysis of Argentinian trade unions, it is no coincidence that only the Peronist government, using its historically strong ties with the labor movement, has been able to force economic changes without the massive worker resistance that other governments have met.

Uganda, the World Bank's model of structural adjustment in Africa, continues to use force to repress political party rallies, political meetings, and pro-democracy forums. While more open, critical journalism is emerging under the Yoweri regime, the latter is tightening its control over what is considered "permissible journalism."

In an example from the Middle East, since the early 1990s, the Egyptian government has used violent Islamic groups targeting the overthrow of the regime as an excuse to circumscribe the activities of all grassroots organizations in the country. The current government is using both legislative and military means to crush radical Islamic opposition to the government and state. These efforts "conveniently" affect all forms of grassroots activism, including social-welfare NGOs dealing with the increasing numbers of poor, and prevent political opposition from organizing or challenging the regime in the wake of SAPS.

The situation in Egypt is reminiscent of the Philippines in the 1980s where the human rights NGOs were negatively affected by the government's campaign to root out and eliminate the communist insurgent army. Social activists, especially NGO workers and human rights lawyers, were particularly targeted by the government because of the government's crackdown on cause-oriented groups believed to be sympathetic to the Communist Party.

While not all regimes have resorted to coercion, many have used a variety of legislative measures in order to ensure economic reform. In Jordan, for example, Moore finds that the entire process of reform has remained highly controlled and directed by the state, leaving little or no room for policy-advocacy by business groups. This process is mirrored in the Tunisian case, where the state has directed economic reforms while reinforcing the strengthening of "traditionalist" Islamic relationships. Loh notes that, in Malaysia, a combination of patronage and highly restricted legislation limiting NGOs, the judiciary, and the press, have been effective tools in ensuring the implementation of reforms. Here, more than anywhere, the mere existence of NGOs is not *ipso facto* a guarantee that they have extensive influence – or that they are a sure sign of democratization. The restriction of political liberalization has ensured economic liberalization. Non-governmental organizations seem to have two significant boundaries in Asia; the amount of strike disruption is limited, and NGOs are limited to political issues.

As Clarke's chapter on the Philippines suggest, human rights NGOs were somewhat successful as long as they focused on issues of human rights violations narrowly defined as civil and political. As soon as they re-defined human rights to include social and economic rights, and entered the realm of criticizing economic restructuring and the loss of "entitlements," their influence waned significantly *vis-à-vis* the state. This partial marginalization has been exacerbated by divisions within the NGO community which were largely caused by the government's co-optation of specific human rights groups and not others. Furthermore, now that the government has created official commissions and boards to deal with human rights abuses, there is declining support (and no receptive avenues) for alternative or deeper definitions of human rights abuses.

Clark's study of Egypt demonstrates how the continued application of Law 32 of 1964, the law regulating all NGOs in the country, prevents NGOs from playing a greater policy-making role by limiting their ability to raise funds and their political autonomy. Similarly in Kenya, Nyang'oro finds that civil society suffers from a distinct lack of autonomy. In Nigeria the military government has arrested NGO and opposition leaders, "castrated" labor unions, "emasculated" the judiciary, and nullified election results.

These examples challenge the assumption that autonomous political space is being created in the wake of structural adjustment. Political restrictions prevent grassroots associations from being able to take advantage of any newfound political space. Dicklitch's commentary on NOCEM in Uganda is a particularly stark example of this. While economic space is being created, political space is not. As a consequence we find the autonomy of civil society from the state in question. In contrast to the neo-conservative understanding of civil society, any distinction between civil society and the state is highly blurred and difficult to make. The state continues to dominate the political arena through legislation or through coercion.

Indeed, authors such as Bensabat Kleinberg, Patroni, Clark and Isbester find that, despite drastic economic changes, semi-authoritarian corporatist structures remain. The majority of governments in this volume seem to be in a process of change but are unable or unwilling to break away from semi-corporatist and corporatist forms of state–society relationships, where coercion may be used as much as co-optation is. The semi-corporatist state structure in Egypt has remained intact as SAPs have been implemented. It continues to be used as an instrument of top-down control.

The Mexican state and its one-party system is also unwilling to totally renegotiate its corporatist ties to civil society at large. As Bensabat Kleinberg demonstrates, it is reluctant to fully democratize either the party or the state as it continues to liberalize the economy at a rapid pace. Traditional political arrangements continue to fuel state–business–labor alliances, but without the security of state-supported entitlements. The state's reluctance to dismantle the once-powerful public workers' unions, and the unwillingness of big capital interests to encourage the breakdown of semi-autonomous, legislated, business associations, has led to a form of *neo*-corporatism. That is, economic liberalism has not encouraged the reorganization of top-down corporatist associations into smaller, autonomous interest groups. Powerful capital interests have found a coherent (and convenient) structure through which to press their demands, while at the same time appearing unified as a class.

Similarly, the Peronist government in Argentina is caught between the twin needs of maintaining the labor movement's political support while at the same time curbing its oppositional capacity. As a result of this dilemma the Menem government has attempted to fracture and weaken the movement short of weakening the corporatist structure that grants the unions power.

King finds a very interesting development in Tunisia. Because structural adjustment created considerable inequalities in Tebourba, state elites deliberately fostered redistribution along neo-traditional lines in order to control social and political instability. The most immediate effect has been the encouragement of "primordial identification" based on blood ties and religion.

In addition, Bensabat Kleinberg, Clark and Moore find that the implementation of economic reform has, in at least one respect, had the opposite effect of its original intentions. In all three cases new ministries have been created – rather than being down-sized or eliminated – to regulate economic reform and the role of private organizations. In other words, political space is being created, yet it is the state which is stepping in to fill that space.

Both Clarke's and Isbester's case studies, respectively, indicate that state strengthening of civil society is not mutually exclusive. The processes can happen simultaneously. In the Philippines the state has been able to introduce economic liberalization, incorporate human rights organizations' demands on issues that do not threaten economic reform, and devolve minimal decision-making. In Nicaragua the state has technically decentralized, yet this process has eliminated women's channels to decision-makers and, therefore, a voice. Economic liberalization has not resulted in greater participation for women's groups.

While civil society may not be strengthening in the wake of economic restructuring, in most cases there is no doubt that there has been a definite shift in the relative balance of associations within civil society. Civil society is not only unequal but divided. Sidney Jones, for example, notes the increased tensions between Indonesians and ethnic Chinese minorities as economic reform has progressed in Indonesia. In the wake of economic reform in Kenya, increased ethnic divisions, erupting upon occasion into violence, have accompanied economic and political change and fragmented civil society. In Nyang'oro's study of multi-party Kenya we see parties using ethnicity as a mobilizing force, thus weakening their ability to forge a united opposition to the ruling party. Years of political repression have created a culture of violence within Kenya's civil society. Political opposition groups have disrupted each others' office elections, rioting is a regular avenue of "participation," and ethnic divisions are growing.

Similarly in Jordan, the historical division between East and West Bankers, the pre- and post-1967 populations, respectively – has dominated the process of economic and political reform in at least two important ways. On the one hand it has divided business groups and circumscribed their effectiveness. On the other hand the monarchy is acutely aware of the fact that its power base lies with the East Bankers. As such, the monarchy has been hesitant to wholeheartedly embark upon privatization for fear that too many important economic assets will come into Palestinian hands.

Sectoral divisions between business have also left their impact in Mexico where big business has been able to make siginificant gains at the expense of medium and small business. Bensabat Kleinberg reinforces the fact that, while there has been a realignment of the strength of various business groups since economic restructuring has occurred, this has not meant greater political liberalization or democratization.

In Malaysia also, big business has forged closer ties with the state in order to further its interests as human rights groups have more vocally criticized the resultant inequalities of economic reform. The Federation of Malaysian Consumers Association, for example, successfully protested against the exorbitant (and unsubstantiated) increase in sewage service rates by a private firm after it was awarded the entire national sewage system as part of the country's structural adjustment program. This protest largely reflected NGOs' concern over the deepening of patronage in the awarding of concessions to private firms and the lack of transparency in the regulation of private monopolies and pricing.

In other instances civil society has been weakened by the creation of government-sponsored organizations (GSOs). In Egypt, community development associations, created by the government, receive the greatest degree of government funding. They tend to have close historical ties with the Ministry of Social Affairs and are often directed by retired senior ministry officials with government-seconded staff. Their ties to the government wed these associations to the *status quo*. Importantly, GSOs also serve to (re)-legitimize the government in the face of decreasing economic entitlements.

The Museveni regime in Uganda has similarly created a rival body to the Ugandan Journalists Association (UJA). The government successfully divided and weakened both the UJA and its opposition to the state. Similarly, the military regime in Nigeria has co-opted opposition leaders to such an extent that it has resulted in the muzzled silence and splintering of opposition groups.

In the Philippines, Corazon Aquinos (1986–92) created the Commission on Human Rights (CHR) and the Presidential Human Rights Commission (PHRC) in 1987 to address the expanding, and broad-based Philippine Alliance of Human Rights Advocates (PAHRA). Tensions between the NGOs and the GSOs led to the reformulation of the PHRC in 1988 and its incorporation of representatives from the NGO community. By the time José Ramos was elected into power in 1992, the human rights situation had been substantially improved but, even so, Ramos took measures to insure linkages between his government and the human rights NGO community. A Fact Finding Committee on Involuntary Missing Persons was created, and a compensation fund was announced. The willingness of NGOs to establish linkages led to "constructive engagement" with the state and hence legitimacy for the latter. At the same time, as Clarke demonstrates, the NGO community suffered its own fragmentation, damaging its own credibility and weakening its base of opposition to the Ramos government.

We, therefore, see a lack of civility or tolerance within civil society – undermining the role of civil society in creating a participatory civic culture. However, are the organizational structures of civic associations participatory? Is there "internal democracy" within civil associations? What seeds for future democracies are they planting?

Each of the case studies exemplifies the infancy of civil society in their respective countries, an infancy that has hindered its ability to act. While civic associations may be well-intentioned, they are often not participatory or democratic in their functioning. Social-welfare NGOs in Egypt exhibit structural weaknesses, a lack of participatory decision-making, a shortage of facilities and training, and little or no coordination between NGOs. Few Philippine human rights NGOs have a sizeable or professionally trained staff. There are furthermore many – perhaps too many – small human rights organizations as groups splinter from larger organizations. Similarly, in Indonesia, the Prosperous Indonesia Workers Union suffered from the fact that it lacked "colonels." When the "general" was arrested there were plenty of "foot soldiers" but no colonels, so the organization came to a halt. Dicklitch also finds that, while state restrictions limited the effectiveness of political parties in Uganda, internal bickering and in-fighting significantly weakened their organizational structure. King found that, as primordial identification increased in Tunisia, membership in organizations representing cross-cutting cleavages declined during economic reform.

In Malaysia, civil society has largely acquiesced to the consolidation of executive power that has accompanied economic reform and privatization. Widespread mass consumerism, individualism and political stability, all driven by economic reform, have become highly valorized in Malay society. As a result, Loh argues, support of the current non-democratic regime appears to be the price Malays are willing to pay for market-driven economic development and its by-products.

Moore's study of business groups in Jordan reveals incoherent business leadership and intensely divided interests between sectors of business. The organizational weaknesses and internal divisions within business leadership ranks tend to reflect larger divisions within the business community at large. Thus, private sector representatives, according to Moore, appear unable to exploit economic (or political) liberalization to the extent that other groups have. While private sector interests in Jordan are concerned with issues of transparency and specific economic accountability, they are far less interested in broader issues of political representation and democratization.

Similarly, Patroni observes, in trade unions in Argentina, repeated efforts by union leadership to eliminate any internal opposition. The Argentinian case is marked by a distinct lack of democratic values. Patroni's examination of Argentina's labor movement reveals that Argentina has not experienced the democratization of either the state or the labor movement. In order to maintain its supremacy within the labor movement the Peronist General Confederation of Labor (CGT) is marked by the exclusionary nature of its leadership toward the representation of alternative interests within the movement. Labor has essentially preserved its authoritarian and hierarchical structure in an effort to suffocate the emergence of more radical and democratic forms of unionism.

Even in the diverse and expansive business environment in Mexico, Bensabat Kleinberg shows that big-business leadership prefers to maintain, rather than dismantle, government-created mandatory business associations. In other words, big business in Mexico continues to prefer corporatist to democratic linkages with the state. Democratization simply does not lie in these groups' interests. For labor in Argentina, and business in Mexico, democratization means a devolution of specifically their power. It is for this reason that Argentina's labor movement and big business in Mexico resist any attempts to dismember the corporatist structure and their ties to the state, even at the expense of political autonomy. Facing an acute tension between economic reform and democratization, democratization has been sacrificed, if not subverted, in these case studies.

Hence, governments' resistance to political liberalization and democratization is often buoyed by elements of civil society. Clearly, not all (or even the majority of) civil associations are anti-democratic, but these case studies are a reminder that civil associations cannot be assumed to be democratic in their decision-making or in their interactions with one another, nor can they be assumed to be necessarily cause-oriented politically. In other words, as Nyang'oro and Dicklitch each point out, many civil associations pursue essentially apolitical goals and would prefer, if possible, to remain apolitical. Clark's example of Egypt is a case in point. Social-welfare NGOs in Egypt explicitly target the immediate needs of the poor – medical care, for example. By and large they have not entered the political arena. There is no doubt that they have avoided political activism due to their lack of autonomy from the state and the government's careful watch over all forms of social activism. However, they have also remained apolitical due to the important, challenging and overwhelming nature of simply doing their job of providing social-welfare services to the increasing numbers of poor. Without maligning these types of associations in any way, many civil associations are simply not *actively* interested in democratization. Similarly, big business in Mexico is simply doing its job, regardless of what it means for democratization and social justice in Mexico as a whole.

Hence, the chapters in this book reveal a broad range of types of civil associations in terms of their goal-orientation, degree of political activism and political skills, and their degree of democratic behavior both within civil associations and between them. These differences affect civil associations' willingness and/or ability to act as agents of political mobilization, political liberalization or democratization. They furthermore affect the nature of civic culture.

The degree of commitment to democratization that civil associations demonstrates and/or their strength as agents of political liberalization and democracy are largely circumscribed by the parameters established by their governments and domestic political dynamics. The case studies in this volume reveal governments whose imperatives can be placed along two continua not wholly visualized by the international financial community. The first spans from greater democratization on the one end, through to greater state intervention and ultimately military control on the other. The second extends from governments' desire to provide social welfare benefits – as a means of legitimization – to the economic pressures for greater privatization at the other end. Unlike the causal link shown in Figure 1, the two are not necessarily mutually reinforcing. Rather, we see a less linear, more tension-filled relationship between the two, as shown in Figure 2. Civil associations must operate within this fictional relationship.

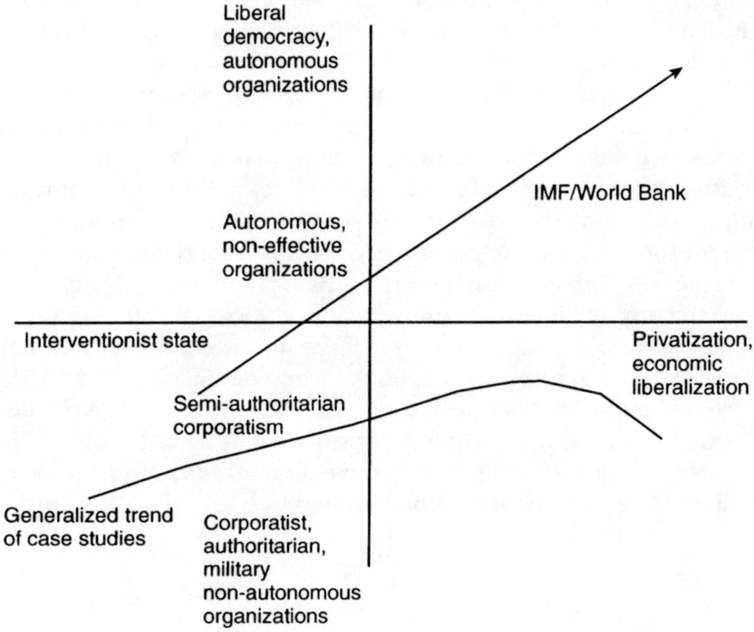

Fig. 2 IFI projections, 1980–99

Regardless of governments' will to democratize, the twin demands of fiscal responsibility (international, financial and political demands), and domestic stability and legitimacy are often at odds with each other. As Figure 2 indicates, while some states seem to be consolidating *institutional* or *electoral* democracy, such as Nicaragua and the Philippines, the majority of governments discussed in this volume have taken conditional steps toward political liberalization but have undergone little or no devolution of decision-making.

The implementation and impact of structural adjustment measures have, in almost all cases, necessitated – at least from the governments' perspective – some form of (coercive) state control over civil society. Heavy-handed attempts to mute opposition to economic reform, however, are risking the erosion of regime legitimacy and increasing the potential for political instability and/or chaos. This is particularly the case when governments have not adequately addressed increasing poverty rates and unemployment.

The associations of civil society both operate within and react to economic reform programs and governments' political strategies – ranging from co-optation to coercion – to ensure political stability and continuity. Economic withdrawal has resulted in the creation of very little, if any, "political space" of which civil associations have been able to take advantage.

Yet the tensions between free-market capitalism, transitional political institutions, and the need for economic and social development, have created some opportunities for civil society. While the general trends highlighted by this book indicate increased authoritarianism in conjunction with economic structural adjustment, these trends adopt different forms. Domestic political dynamics create both obstacles and opportunities for civil associations. This Introduction strongly challenges the unproblematic assumption of civil society's ability to further political liberalization and democratization in the context of economic reform, yet the chapters in this book clearly delineate civil societies' "successes" as well as "failures". Civil societies' activities have not been without positive results. The question that will be addressed in the Conclusions to this volume is under what conditions *can* civil society play a role in greater political liberalization?

Notes

1 See Adam B. Seligman, *The Idea of Civil Society* (New York: Free Press, 1992), pp. 10–11.
2 See Steven M. Delue, *Political Thinking, Political Theory, and Civil Society* (Boston, MA: Allyn & Bacon, 1997), pp. 1–7.
3 Michael Bratton, "Beyond the State: Civil Society and Associational Life in Africa," *World Politics*, XLI: 3 (1989), 417.
4 Delue, op. cit., p. 278.
5 Economic reforms here mean the "commitment to improved policy environments in terms of domestic macroeconomic management and economic liberalisation: open door policies toward foreign trade and foreign investment, which contributed to closer integration with the global economy, strong growth of the service sector and expanding trade in services; a decreasing share of central government expenditures." World Bank Notes on the Economy, and State and Markets, World Bank Website.
6 USAID Strategic Plan, Website: http//www.info.usaid.gov/regions/lac/sesd/econ.htm
7 Ibid.
8 Laura Macdonald, *Supporting Civil Society: The Political Role of Non-governmental Organizations in Central America* (New York: Macmillan and St Martin's Press, 1997), p. 14.
9 Saad Eddin Ibrahim, "Crises, Elites and Democratization in the Arab World," *Middle East Journal*, 47: 2 (Spring 1993), 302–5.
10 Julius O. Ihonvbere, "The World Bank/IMF Structural Adjustment Programs and the Somali Crisis." Paper prepared for the symposium on "Towards Conflict Resolution in the Horn of Africa," organized by the African Studies Program, Central Connecticut State University, New Britain, Connecticut, 19 November 1994.
11 William C. Smith, Carlos H. Acuña and Eduardo A. Gamarra (eds), *Democracy, Markets, and Structural Reforms in Latin America* (Florida: University of Miami: North–South Center Press, 1994), p. v.

Part 1
Asia

1

Structural Adjustment and Democratization: the Case of Human Rights Organizations in Indonesia

Sidney Jones

Introduction

In the wake of the dramatic fall of President Soeharto in May 1998, the International Monetary Fund (IMF) and global capitalism were widely given twin billing as the agents most responsible for political change. Globalization meant that Indonesia was unable to escape the impact of the Asian economic crisis after the Thai currency collapsed in July 1997; its authoritarian political structure made it unable to mount an adequate response. According to this view, the successive IMF packages, in October 1997, January 1998 and April 1998, and the structural adjustment these packages required, fueled the social unrest that led the army and the middle class to decide that their only salvation lay in the removal of Soeharto and a process of political reform.

The truth is naturally more complicated. There is no question that by early 1998 the impact of the Asia economic crisis was causing hardship across the Indonesian archipelago. The rise in prices of basic goods sparked a series of riots across the country, targeted at the ethnic Chinese shopkeepers who by some estimates control more than 70 percent of the retail economy. But structural adjustment, in the form of the IMF packages, was not responsible for those riots or price increases because the government had stalled on implementing the reforms. It was Indonesia's currency collapse, unrelated to any program imposed by international financial institutions, that triggered the price rises and layoffs, and it was slow-burning anger at President Soeharto and his family that made his "election" in March to a seventh five-year term unacceptable to much of the population. The convergence was explosive, but structural adjustment had little to do with it.

Indeed, beginning in the late 1960s and early 1970s, it was economic development, not economic restructuring *per se*, that produced increasing levels of education and income, and spurred the growth of an educated middle class. That middle class in turn provided a pool from which non-governmental organizations (NGO) leaders since the late 1970s have been drawn; for the past two decades those leaders have been largely college-educated professionals. Economic restructuring led in some cases to tighter political controls while at the same time producing new targets for NGO activism.

But it would be difficult to argue that either economic growth or structural adjustment led to broad-based demand for systemic political change. Outside the NGO community, and even to some degree within it, the forces for democratization were weak until the full force of the 1997–98 currency crisis hit home. That crisis put the corruption of the first family and the shortcomings of a closed political system into the spotlight. Until then the middle class had not, by and large, become a vocal proponent of political change, although dislike of the Soeharto family was widespread, as was a desire for less arbitrary governance.[1]

Many middle-class Indonesians were wary of democratization, fearing either that it would lead to greater communal differences in a multiethnic state or to a decline in status or standard of living for those who benefit from a highly centralized patronage network. The multiple fault lines of the middle class – "indigenous" Indonesians vs ethnic Chinese, Muslim vs Christian, Javanese vs other ethnicities – all militated against a class-based solidarity. It is true that two decades of activism produced a new militancy among NGOs and student activists but it was a militancy derived less from the frustrations of a disaffected middle class than from the anger of the poor and powerless that NGO activists sought to represent.

This chapter focuses on the impact of economic restructuring on Indonesian labor rights activism as one subset of the broader non-governmental movement. It argues that the growth in export-oriented manufacturing, a direct consequence of the restructuring in the mid-1980s, took place at a time when Indonesian NGOs were growing rapidly in numbers and sophistication and when their links to international NGOs were expanding. A potentially powerful alliance developed in the 1990s when key NGO leaders decided that a restive labor force in the apparel, footwear, and textile industries could be mobilized into a national labor movement and become an engine of democratic change. Many human rights and community development organizations turned their attention to organizing workers and helping them press for higher wages, greater benefits, and independent unions. By 1998, however, the alliance was weaker than it had been four years earlier, the relative economic strength of the industries in question was declining, and alternative "engines" for democratization remained elusive. It was only in the first five months of 1998 that the economic crisis drew students and NGOs, the army, and the middle class into a tactical alliance to bring down the Soeharto government and open up possibilities for democratic change.

Political and economic background

Indonesia, a sprawling archipelago of some 13 000 islands and an estimated 250 ethnic groups, is now the world's fourth-largest country with a population estimated at about 200 million. Once called the Dutch East Indies, it emerged from its colonial past after a bitter and protracted war of independence from 1945 to 1949 against the Dutch, assisted by their wartime allies. Sukarno, leader of Indonesia's nationalist movement, became the country's first president. From 1950 to 1957 Indonesia was a parliamentary democracy but a chaotic one, with six different cabinets. In the 1955 elections, the only free elections the country has ever had, over 100 different political parties took part, many of them communally based. By 1959 Sukarno, disillusioned with what he called "democratic liberalism," had instituted "guided democracy" in its place, with a strong president, an appointed assembly, and a strong role for the army. Political parties were not immediately dissolved, but their role changed from competing in elections to mobilizing their followers to support or resist national policies. Sukarno, free from any parliamentary checks, embarked on a series of politically and economically ruinous policies in the name of struggling against Western imperialism and "standing on one's own feet." He banned ethnic Chinese from trading in rural areas, expelled the Dutch who had continued to own and manage many plantations and industrial projects after the war, embarked on a military confrontation with neighboring Malaysia, and launched a move to wrest West Irian, the western half of the island of New Guinea, from the Dutch. In all these moves he relied on the support of the army and the Indonesian Communist Party (Partai Komunis Indonesia, PKI) which by the mid-1960s was the third-largest Communist Party in the world after those in China and the Soviet Union. These twin sources of support bitterly distrusted each other, however, and as Sukarno's health began failing in 1964, political tension between the two increased. That tension culminated in the assassination of six army officers on the night of 30 September 1965, in what the current Indonesian government has termed an attempted coup by the PKI. The full story of who, in fact, was responsible remains to be told, but the consequences were clear. A Jakarta-based officer named Soeharto mounted a successful "counter-coup," forced Sukarno to ban the PKI, and used the army to encourage the mass slaughter of hundreds of thousands of suspected supporters of the PKI. Some of the worst outbreaks of killings on Java were led by Muslim groups in rural areas who were angered by the PKI's commitment to land reform and saw the destruction of the PKI as a victory for Islam over atheism.

In March 1969 Soeharto formally took power from Sukarno and declared his government to be a "New Order." At the time, inflation was running at close to 1000 percent, there was virtually no investment, the central government had lost much of its authority over the outer islands where most of the country's exports were produced, public administration collapsed, levels of health and education were declining, and there seemed little hope for recovery. Soeharto moved quickly to put in place a highly skilled team of economic managers to try to set the economy to right. These technocrats succeeded beyond anyone's imagination.

Hal Hill, an economist at Australian National University, has roughly divided the economic changes that resulted into four periods: 1966–70, rehabilitation and recovery; 1971–81, rapid growth fueled by the quadrupling of oil prices in 1973, accompanied by the rise of economic nationalism; 1982–86, adjustment to lower oil prices accompanied by much slower growth and an increase in trade barriers; and 1987–present, steady liberalization and a spectacular growth in manufacturing exports.[2]

The 1966–70 period saw inflation brought under control, an annual economic growth rate of over 6 percent, reaching 10.9 percent in 1968, a surge in investment, both foreign and domestic, and rehabilitation of existing physical infrastructure. The US-trained technocrats, known as the "Berkeley Mafia," earned international respect for their success in coping with Sukarno's legacy, but it is important to remember that at the same time the economy was being brought under control, one of the worst massacres of the twentieth century was continuing. In addition to the killings, more than one million people were arrested and detained for suspected links to the PKI. No study has ever been conducted of the economic impact of these events, or whether economic decision-making was hampered or fostered by the terror, which lasted well into 1967–68.[3] It is important, however, to put Indonesia's economic "recovery" in context. The terror may have worked to Indonesia's advantage, as the crackdown on suspected Communists, at the height of the Cold War, may have bolstered Western and Japanese support.

The period from 1971 to 1981 was one of rapid economic growth, but it was also a period of political upheaval and unprecedented strengthening of state authority.[4] Economic policies of the immediate corrective period after 1966 had led to red-carpet treatment for foreign investors. In early 1974, domestic reaction against the ubiquitous presence of Japanese investors set in, after a riot in January triggered by the visit of then Japanese Prime Minister Tanaka to Jakarta. Part of the reaction was that the new investors often had ethnic Chinese as their partners, to the intense resentment of *pribumi,* or indigenous and usually Muslim entrepreneurs. A second was that Muslim students, who had been instrumental in mobilizing support for then General Soeharto, felt marginalized after it became clear that support had brought no obvious benefits to the Muslim political elite. A power struggle within the elite took advantage of these reactions, and different factions helped foment the riot. The successful handling of the tension by Soeharto, combined with the unprecedented flow of resources after the skyrocketing of oil prices in 1973, enabled him to cement his authority as never before, but there were still problems ahead.

The riches flowing from oil revenues had positive and negative results. A substantial amount of the new wealth was poured back into education and other essential investments, but the income also gave rise to a new economic nationalism and a sense that now was the time to divert Indonesia's resources to the *pribumi* elite – which often translated into the family of President Soeharto and their cronies – and away from the ethnic Chinese. One way of doing this was to invest heavily in state enterprises, such as cement, steel and fertilizer. The result was high but highly inefficient growth and increasing concern about corruption. A major scandal in the state oil company, Pertamina, revealed that its debts totaled $10 billion, nearly a third of the country's entire GDP.

That crisis was weathered, but the cronyism and corruption it uncovered helped spark increased student activism in 1977–78 that in turn led to a new wave of repression. Student manifestoes issued in October 1977 and the months that followed focused heavily on the way in which economic development was benefiting Soeharto and his family and hurting ordinary people. As one trenchant student critique noted, "The government is *only* interested in achieving a bigger GNP and in the process has forgotten everything outside that target. It has forgotten the principles of social justice, self-respect, national culture [...] 70 percent of foreign capital is invested in the city of Jakarta. The other 30 percent is spread out all over Indonesia. Where is the social justice in that?" Regional disparities were in fact to become a key political grievance that helped foster the notion of "empowerment" as the NGO movement grew in the 1980s; the roots, however, lay in policies formulated a decade earlier. The student manifestoes also questioned the amount of government credit going to ethnic Chinese as opposed to "the people" and said with regard to law that the prevailing principle was "Legal sanctions for ordinary people, legal protection for bigshots."[5]

But the major impact of the 1971–81 period in addition to economic growth was the concentration of economic and political power in the hands of President Soeharto and his closest associates. The economic power was achieved through trading monopolies, heavy controls on investment, and erection of high-tariff and non-tariff barriers; imports were regulated by a system of "approved traders" who often had ties to the Soeharto family. The political control was achieved by the forced merger of Indonesia's various political parties into three, and by the operation of KOPKAMTIB, or Command for the Restoration of Order and Security, an internal security agency controlled by military intelligence that extended its search for political enemies beyond suspected communists to suspected Muslim militants, particularly in the wake of the Iranian revolution. The invasion of East Timor, in December 1975, also took place as economic growth was soaring, and Indonesia was increasingly regarded as a model of Third World development. The lack of international reaction to the invasion, however, had more to do with the fall of Saigon and Indonesia's regional role as an anti-communist bulwark. (The lack of domestic reaction was due to tight government controls on information, total military control of the territory, and no interaction to speak of between non-military Indonesians and East Timorese.)

It was during this period that some of the best-known NGOs in Indonesia were established, a result in part of the perceived neglect of the "little people" by the New Order. These included the Legal Aid Institute in 1970, set up to provide legal aid for the poor and work for adherence to the rule of law, and the Indonesian Environmental Forum, known by the acronym WALHI, in 1980.

The 1982–6 period of adjustment to lower oil prices and efforts to diversify the economy had effects that were to be far-reaching with regard to the NGO movement, particularly in terms of the focus on light export-oriented manufacturing. The immediate impact, however, was a realization on the part of government technocrats that the oil revenues of the 1970s had served to subsidize highly inefficient producers, and the task they set for themselves was to lower high production costs.[6] They were only partially successful. The continued fear of ethnic Chinese dominance and the benefits that accrued to Soeharto cronies from the heavy role of the state were powerful barriers against reform. Indeed, trade policy came to be seen as an instrument of industrial policy, ensuring that associates of the family and other key members of the political elite had control over imports seen as essential to the production needs of state enterprises and family-owned businesses.[7]

The period 1982–86 was also a period of consolidation of President Soeharto's control. In 1979–80, a challenge to Soeharto had emerged from within the ranks of respected army officers and other former officials, the so-called Group of 50 or *Petisi* 50. In part to respond to this challenge, in part to preempt the growth of more fundamentalist Muslim groups that drew inspiration from the Iranian revolution, Soeharto pushed for ideological uniformity through two means. One of these was adoption of the state ideology, Pancasila, as the sole ideology of all mass organizations, and the institution of Pancasila training courses. Pancasila, a word meaning "five articles," was, on the surface, a set of unobjectionable principles that Sukarno had borrowed from Indian Prime Minister Jahawaral Nehru. They were belief in one God; humanitarianism; nationalism; democracy; and social justice. Under Soeharto, adherence to Pancasila became equated with loyalty to the New Order government. Resentment against the policy of *asas tunggal* or "sole foundation" policy therefore ran deep, especially among religiously affiliated organizations who believed Islam or Christianity had been and should continue to be their ideological foundation, and it was a major factor in the outbreak of serious riots in the Tanjung Priok area of northern Jakarta in September 1984. A draconian anti-crime law in which the police and army were given authority to summarily execute suspected members of criminal gangs without charge or trial gave rise to an estimated 4000 killings of young men between 1983 and 1985 – the so-called *petrus* or "mysterious killings" since the perpetrator was never identified and the murders never investigated. With the *petrus* killings the government could both respond to middle-class fears of escalating crime and send an unambiguous message about the extent to which it would go to ensure that "development" could proceed unhindered.

Economic restructuring in Indonesia

The impetus for structural adjustment in Indonesia was the fall in oil prices that reached bottom in 1986 but, as one analyst noted, "The oil crisis was not a single event that made 'shock treatment' necessary but rather an extended crisis that made gradualism possible."[8] The adjustment was also incomplete, particularly in the area of privatization. Structural adjustment is usually defined as an economic package consisting of trade liberalization through removal of tariff and non-tariff barriers, privatization of state enterprises, encouragement of foreign investment, and cuts in state spending, including removal of subsidies of basic goods.

The shock of the 1985 oil price drop proved to be the impetus for an effort to implement many of the elements of a structural adjustment package in order to diversify the economy and spur the growth of non-oil exports. Awareness of the need to revamp the economy was already evident, as the fourth five-year plan (1985–9) had given unprecedented prominence to the private sector. A series of deregulation packages were announced beginning in October 1988, and the next two years saw major steps toward the deregulation of banking. For some NGOs working in rural areas, deregulation was a major boon because it meant easier availability of rural credit to foster income-generating projects and local savings schemes and, ultimately, closer cooperation with the government. For example, one of the longest-established and most professional of the NGOs, a group called Bina Desa (literally, "village development"), was able to set up a savings and credit program called "Pilot Project for the Development of Cooperation between Banks and Community Self-Support Groups." By 1994 over 1000 village-level institutions for dispensing small loans were in operation.[9]

In terms of economic performance alone, the deregulation package worked. Between 1988 and 1991 the economy grew by about 9 percent per year. By early 1992 non-oil exports and foreign direct investment were on the increase, with much of the growth in the export-oriented manufacturing sector, particularly textiles, clothes and footwear. Those three products accounted for almost half of all jobs created in the large and medium manufacturing sector between 1986 and 1992, and in terms of the value of exports the three grew at a rate of over 15 percent between 1988 and 1991.[10]

Three more important deregulation measures were announced in May 1995, January 1996 and June 1996, all aimed at reducing tariffs and facilitating imports for use in production of exports by foreign firms. Little progress was made in reducing the role of state enterprises in production of essential or strategic goods; indeed, the rise of B. J. Habibie as Minister of Research and Technology and head of the State Agency for the Assessment and Application of Technology gave state enterprises a new lease of life. The reforms also made little dent on the extent of state subsidies for basic needs, in part because of the immediate popular protest that followed the announcement of any reduction or concomitant rise in prices. Street protests over a hike in electricity prices in 1989 and in public transport fares in 1996 were two that led to major crackdowns.

Overall, the impact of economic restructuring was a high growth rate and a successful shift away from oil to non-oil exports; not only textiles, garments and footwear, but also pulp and paper and wood products. (By 1995–96 chemical production had overtaken both.) Foreign direct investment (FDI) was at an all-time high by 1996, with approvals for FDI projects having risen from $1.502 billion in 1987 to $39.915 billion in 1995.[11]

On the negative side, restructuring fostered the growth of a class of ostentatious new rich, the wealthiest of all being the Soeharto children and their ethnic Chinese business partners. Large conglomerates flourished, driving underground small shops and enterprises. Public perceptions of a rapidly widening income distribution gap were at first not borne out by economic data; by 1995 Indonesian economists were finding that facts were catching up to perceptions, and the gap was indeed growing larger. The resentment and "social jealousies" (*kesenjangan sosial*) spawned by the gap were producing major social unrest by late 1996, not only in Java, the traditional site of anti-Chinese violence, but also in Kalimantan where the growth of palm-oil plantations, tree plantations for pulp and paper factories and other commercial enterprises was bringing in transmigrants from Java, Madura and Bali and systematically dispossessing local people of their land.

The labor movement

Developments within the labor movement in Indonesia perhaps represent the clearest link between economic restructuring and demands for political change. By the mid-1990s something resembling an independent labor movement had emerged for the first time in three decades, the result of the growth of export-oriented industry and the increased strength of NGOs. It was clear that student activists and some intellectuals saw an alliance with labor, and the mobilization of disgruntled factory workers, as central to democratic change, although as of mid-1997 it remained to be seen whether workers across industrial sectors would themselves find common interests in a way that could exert significant political pressure on the Soeharto government.

Any independent labor movement in Indonesia has a significant historical burden to contend with. In the 1950s and 1960s the Indonesian trade union movement was alive and well, with labor federations identified with each of the major political parties. By 1954 SOBSI, the federation affiliated to the PKI, had become the country's largest, and by 1965 it claimed a membership of 3.5 million, between 50 and 60 percent of all Indonesian union members. The PKI leadership estimated at the time that SOBSI members and their families constituted a full 10 percent of the entire Indonesian population, then some 115 million.[12] (In 1992, by contrast, the Indonesian government stated that the single government-controlled federation, the SPSI or All-Indonesia Workers Union, had a total membership of about 1 million, and the population by then had grown to 190 million.)

Following the 1965–66 killings, the ban on the PKI, and the coming to power of the New Order government, the labor movement was crushed. The dominance of SOBSI led the Soeharto government to make a permanent association between labor organizing and communism, and as late as 1992 an Indonesian government statement noted, "One possible inroad for communist subversion is through worker unrest. The Government especially fears that the communist ideology might be attractive to young people who did not live through the national distress of the 1960s, and many of those young people can best be reached on the job through labour disputes."[13]

Strict controls were therefore placed on unions. The New Order government forced the 13 party-affiliated unions to merge into a single Consultative Council for Indonesian Labor which in 1973 became the All-Indonesian Labor Federation (FBSI), directly under the control of the Ministry of Manpower – not Labor Ministry, as that had a leftist ring. Other federations were not formally banned, but the requirements for registering were so onerous as to make any challenge to the FBSI impossible. The FBSI did almost no organizing; one researcher reported that, in the factories she was studying in Central Java, the

FBSI representative did nothing but organize the occasional volleyball or table-tennis game.[14] In 1985, after an outbreak of industrial unrest, the federation was restructured to tighten government controls and further reduce its independence. As a single union in form now rather than a federation, its initials changed from FBSI to SPSI, but if anything, it was less of a workers' organization than its predecessor, with government officials and army officers rather than workers making up its management structure and little interest in organizing or collective bargaining.

For the first decade and a half of the New Order there were very few strikes that reached the government's attention or found their way into the statistics of the ILO, in part no doubt, because of the very real danger that anyone who tried to organize a work stoppage could be labeled a communist, in part because much Indonesian manufacturing was confined to cottage industries, in part perhaps because of stricter controls on the press than existed a decade later. For whatever reason, from 1966 to 1975 only seven strikes were recorded nationwide.[15] No one was championing the cause of workers; the 1978 student manifestoes which one scholar called "the first systematic Indonesian critique of the domestic policies of the New Order regime," barely mentioned labor.[16] Still, a trickle of complaints from workers began coming in to organizations such as the Legal Aid Institute. In 1980 the Institute recorded some 30 strikes, some of them brief stoppages in solidarity with dismissed workers but over 20 centering on wage increases.[17]

Following a 1978 devaluation and cutbacks in industries nationwide, the number of strikes picked up sharply, many of them in connection with layoffs of workers, to the point that the government stepped in not only with the tighter restrictions on organizing but also with a ministerial decree authorizing military intervention in labor disputes. Between 1986 and 1990 government statistics showed the number of strikes per year falling off sharply, although there was undoubtedly much under-reporting. But industrial unrest increased dramatically thereafter, to the point that, by 1994, government data put the number of strikes at 1130, over 90 percent of them in manufacturing.[18] The number of strikes had not only risen steadily but each action was also involving far more workers than before: at the Gajah Tunggal textile factory in 1991 some 14 000 workers went on strike.

Why the sudden increase? Local activists attributed it to the deregulation packages of the late 1980s and early 1990s that encouraged foreign investment and the relocation to Indonesia of companies and contractors mostly from Korea, Taiwan and Hong Kong, who acted as outsourcers for major international corporations. This phenomenon in turn had several consequences. First, major wage discrepancies emerged between the Indonesian workers and the foreign, often Korean or Taiwanese, managers. One 1992 study by Akatiga, an NGO in Bandung, West Java, of shoe factories showed the wage ratio between Korean manager and the mostly female Indonesian workers to be 220 to 1.[19]

Second, the factories of the export-oriented industries in general employed far more workers on the plant grounds than other industries, and the factories themselves were clustered in specific geographic regions, mostly around Jakarta, Bogor, Tangerang and Bekasi in West Java (usually abbreviated JABOTABEK after the first few letters of each town), but also in Bandung, West Java; Surabaya, East Java; and Medan, North Sumatra. The number of factories seemed to multiply overnight. In 1986 Indonesia had had ten footwear factories; the number had grown to 141 just three years later, with each factory employing anywhere from 800 to 7000 workers.[20] It was no coincidence that in 1991, according to one study, 74 percent of all strikes in West Java were in textile and garment industries, while 12 percent were in the footwear industry.[21]

Worker discontent was fueled inadvertently by the government itself through the widespread publicity in 1989 over the efforts to regularize a minimum wage law; the Minister of Manpower even declared 1990 the "year of wages."[22] Many factories did not comply, or sought to cut back on other allowances to compensate, and the workers decided to fight back. Periodic increases of the minimum wage thereafter only served to highlight the uneven implementation and lack of adequate enforcement of the law.

Finally, and related to the issues above, labor rose much higher on the agenda of both NGOs and the small Indonesian dissident community, and NGO activists began doing active organizing in factories, particularly in Java and Sumatra. In 1990 a group of activists led by the veteran human rights campaigner H. J. Princen and his Institute for the Defence of Human Rights founded an independent "union" which they named Setiakawan, or Solidarity after the Polish organization. It held training sessions for workers to educate them about their rights, but was never really able to organize enough people to take off as a movement.

Two years later a more ambitious initiative was undertaken by a group of workers dissatisfied with Setiakawan, supported by key members of Jakarta's liberal community. Prominent among the latter was Abdurrachman Wahid, a leading Muslim intellectual and head of the 40-million-member Muslim organization called "Nadlatul Ulama"; Sabam Sirait of the Indonesian Democratic Party (Partai Demokrasi Indonesia or PDI, one of two small officially approved opposition parties); and Asmara Nababan, then head of a labor rights NGO and later to become a member of the Indonesian Human Rights Commission. On 26 April 1992, after a three-day "national workers meeting" in Cipayung, West Java, involving over 100 people from 18 provinces, they announced the formation of the Prosperous Indonesia Workers' Union (Serikat Buruh Sejahtera Indonesia or SBSI), with a labor lawyer and NGO activist from North Sumatra named Muchtar Pakpahan as general chairman.

In terms of the key question of this book – the relationship between structural adjustment and democratization – the formation of SBSI was a major development because it was an indication that the leading advocates of pluralist democracy in Indonesia saw both the extent of worker discontent and the potential for using it as a mobilizing tool of not-so-subtle opposition politics. Abdurrachman Wahid, the Muslim leader, was the founder of Forum Demokrasi, a Jakarta-based grouping of the dissident elite formed in 1991; he was also a long-term campaigner for democratic pluralism. Sabam Sirait was one of those within the PDI trying to turn the party into more than just a loyal opposition. (The timing of the SBSI meeting, just six weeks before national parliamentary elections on 9 June 1992, was significant.) The convening of the "national workers' meeting" was a bold step in and of itself as a test of the openness of the political system; the announcement of a new workers' federation was more than a test – it was a direct challenge to government institutions.

Once the Cipayung meeting was over, the work of organizing was left to Pakpahan, and he started in the JABOTABEK area where most of the export-oriented firms were located. In a speech to install the newly elected officers of the JABOTABEK branch of the SBSI on 19 July 1992, Pakpahan noted that the SBSI had been formed because of growing problems with arbitrary dismissals of workers by government officials, inadequate wages, unclear guidelines on overtime, no social security, failure to provide vacation time, and widespread violations of regulations on hours and working conditions. While the venue of his speech indicated particular attention to workers in the garment and footwear industries, he called on workers from other sectors to join – those in state enterprises, transportation, small-scale trade, teaching and other occupations. He also called on workers to join forces and work together with all parties, with particular attention to the press, NGOs, and political parties.[23]

It turned out, as NGOs had already discovered, that mobilizing workers was easy, especially over the question of wages and holiday bonuses. By January 1993, less than a year after the SBSI was formed, it claimed to have 48 branches in 21 provinces and enough members to meet the requirements for legally registering a trade union, although there was no independent verification of the claim.[24]

Mobilizing international support for a nascent workers' movement was even easier. Until the appearance of SBSI, there had been no real alternative to the official union that international labor bodies such as the International Confederation of Free Trade Unions (ICFTU) could back. Once the SBSI came into being it provided a focal point for international campaigning for freedom of association in Indonesia, with the AFL-CIO in the US and the Dutch federation FNV taking a particularly prominent role, in addition to the ICFTU.

Pakpahan and his associates had a number of factors in their favor that helped raise the profile of SBSI. They had good ties to NGOs for the most part and could piggyback off ongoing organizing efforts. They had a flair for publicity and managed to get virtually every incident of harassment of SBSI organizers into the domestic and international press. (While the government did not formally ban the organization, security forces routinely broke up organizing meetings, often detaining participants, and encouraged employers to dismiss SBSI members.)

They also benefited from developments that they could not have foreseen. SBSI was established just two months before the US Trade Representative agreed to undertake a review of Indonesia's labor rights practices under the Generalized System of Preferences (GSP) provisions of US trade law, effectively requiring the US to cut tariff privileges to developing countries that are not taking steps to improve internationally recognized labor rights. The GSP review was focused on freedom of association, and the SBSI's bid for recognition became an important test case. With his impeccable sense of timing, Pakpahan managed to organize meetings or rallies, which the Indonesian government duly broke up, days before key decisions in the GSP process were scheduled, ensuring maximum publicity about repression of the labor movement.

The SBSI also benefited from the publicity surrounding the murder in June 1993 of Marsinah, a young labor organizer in a watch factory in East Java, who was killed, apparently with the involvement of the local military, after trying to intervene on behalf of her fellow-workers who were detained after they went on strike. The murder set off an outcry throughout Indonesia and galvanized the activist community, especially students and Java-based NGOs, to mobilize around labor issues more intensively than ever before. New networks of activists came into being, such as the Committee for Solidarity for Marsinah, linking NGOs in Surabaya (East Java), Yogyakarta (Central Java) and Jakarta. The murder served to raise public consciousness about labor rights issues to a new level.

The beginning of 1994 saw the announcement of several legal reforms by the government and a rash of factory strikes, mostly in Java and Sumatra. The majority of these strikes were over demands for higher wages and a holiday bonus as the end of the Muslim fasting month, Ramadan, approached. One NGO calculated that in East Java, out of 42 strikes in February, 32 were over the holiday bonus. In Medan, North Sumatra, where several different NGOs had been organizing export industry workers in the Belawan industrial estate, a series of wildcat strikes took place during March, involving several thousand workers, one of which turned violent and resulted in the death of a worker. (Although the factories in the Belawan estate and others involved in industrial action produced goods for export, none was in the textile, garment or footwear industries.) Several weeks later, on 14 April, some 15 000 workers turned out in the streets for a major rally to press for higher wages and demand an investigation into the earlier death. When the governor of North Sumatra failed to meet with the workers, they went on a rampage, vandalizing shops and banks, some of them owned by ethnic Chinese. One ethnic Chinese factory owner died from a heart attack after he was dragged from his car by angry workers. Security forces moved in, arresting hundreds of workers, over 80 of whom were later brought to trial. Muchtar Pakpahan and the SBSI claimed credit for the rally, but in fact it had been organized by a number of Medan NGOs with the SBSI initially playing only a small role. Officials of the Medan branch of SBSI were arrested almost immediately; Pakpahan himself was formally declared a suspect in June and arrested in August. With the top leadership incarcerated, SBSI went into a decline from which it never really recovered, and the labor movement itself seemed to go into a deep freeze, with NGO activists lying low.

The temporary chill, however, led to an analysis among NGOs of why SBSI had been so easily paralyzed. One conclusion, according an activist who attended a meeting of major Jakarta-based NGOs in January 1995, was that SBSI "had a general and lots of soldiers but no colonels." Once Pakpahan, the "general," was arrested, the organizing came to a halt. As these NGOs were organizing a legal aid program for those arrested and trying to come to terms with the Medan crackdown, a much more militant group had emerged in late 1994 among student activists belonging to a newly formed organization called the People's Democratic Alliance (Persatuan Rakyat Demokratik or PRD). Based in the university town of Yogyakarta, but including students from elsewhere in Java and in Menado, the capital of north Sulawesi, these students saw the labor movement in much more ideological terms. They faulted Pakpahan for being too moderate and too focused on educating workers instead of getting them out in the streets. He also, they felt, had the wrong strategy. Instead of starting in Jakarta and building a top-heavy structure in the capital, he should have ensured a steady stream of industrial actions in other cities and only later moved to the capital.

Dissatisfaction with SBSI led to the birth of the Center for Indonesian Workers' Struggle (Pusat Perjuangan Buruh Indonesia or PPBI) with a young woman named Dita Sari, a law graduate of the University of Indonesia, as head. Highly politicized and steeped in banned leftist literature, PPBI leaders decided to organize strikes in factories that, in addition to having dissatisfied and poorly paid workers, also had both political connections to senior government officials or the first family and contracting relationships with multinational corporations. In PPBI-organized strikes the slogans were as likely to be against "militarism" as in favor of higher wages.

In June 1995 PPBI helped organise 13 000 workers at the Great River factory in Bogor, West Java. Great River was a garment company producing Arrow and Kenzo shirts that was partly owned by the Minister of Manpower and a former security minister, Admiral Sudomo. In December, the PPBI got 14 000 workers out in the streets from the Sritex factory in Solo, central Java, a textile firm in which President Soeharto's daughter was a major shareholder. In Jakarta, in mid-June 1996, PPBI helped 5000 workers from several footwear companies hold a rally outside the national parliament building to demand not just higher wages but their own political party and an end to state oppression.

The PRD had formally changed from an "alliance" (*persatuan*) for popular democracy to a "party" (*partai*) on 15 April 1996 with the PPBI as its labor affiliate. The new organization was militantly pro-democracy, calling for an end to Indonesian government control over political parties; a referendum in East Timor on independence; support for Megawati Sukarnoputri, daughter of the late president Sukarno and head of the PDI, for president; international monitoring of the 1997 parliamentary elections; and an increase in the minimum wage. At the same time PRD decided to join forces with other NGOs, student groups, and "people's organizations" in an alliance for democratic change. One of those NGOs was a coalition called the Indonesian People's Council (Majelis Rakyat Indonesia) that included SBSI and was headed by Muchtar Pakpahan.[25] Pakpahan had been released from prison on a technicality in November 1995 and immediately immersed himself in the pro-democracy movement that centered around support for Megawati. After the government, fearful of the latter's growing popularity, engineered her demise as PDI head in June 1996, her supporters occupied the PDI headquarters in Jakarta until July, when government-backed thugs stormed the building. The action triggered the worst riots in Jakarta in two decades, and virtually all of the PRD leadership, as well as Muchtar Pakpahan, were arrested. The students were later sentenced to heavy terms; the trial of Pakpahan on subversion charges dragged on in Jakarta through early 1998.[26]

Many NGOs in urban areas continued to place labor rights at the top of their agenda, with the most active including a coalition of Indonesian and international NGOs called the International NGO Forum on Indonesian Development or INFID; a church-based organization called Yakoma; the Indonesian Legal Aid Institute; and the Jakarta Social Institute. But by 1997 the combined effect of arrests in 1994 and 1996 had been to greatly diminish the role of NGOs in organizing workers. NGOs with labor on their agenda shifted their attention to the international anti-Nike campaign and other company-specific or industry-specific action where international support was readily available. The effect was to make international brand-name companies like Nike the villain, rather than the Indonesian government. At a time when the footwear industry was already on the decline in terms of share of Indonesian exports, their result was also to bury the broader democratization objective of the labor–NGO alliance in favor of the goal of trying to force Nike to pay higher wages.

The fall of Soeharto may produce the real test of how strong the labor movement really is. Muchtar Pakpahan was one of the first two political prisoners released by Soeharto's successor, B. J. Habibie, who also announced that there would be no further obstacles to SBSI organizing. At the same time, as controls on political parties were also lifted in May, the more active leaders of the official union, the SPSI, announced that they were forming the Indonesian Worker's Party to compete in the next parliamentary elections. With unemployment at record levels and demands for higher wages undercut by the clear inability of many companies to even meet current payroll levels, it will be interesting to see whether the workers can make themselves heard in a more open system.

Conclusion

Structural adjustment in Indonesia, by facilitating the growth of export-oriented industries, created the conditions for the emergence of a labor movement. Thousands upon thousands of poorly paid factory workers, many with a high-school education, working in close proximity to one another in industries that had strong international connections, could well have been a catalyst for political change. Over a period of about a decade, from the mid-1980s to the present, the plight of workers increasingly became the target of NGO and student activism, and from 1992 to 1994 it appeared as though a genuine movement based on worker solidarity across regions and even across sectors might actually take off. That it did not was due to several factors. One was the peculiar combination of state repression and state concessions. The crackdowns following the riots in Medan in 1994 and Jakarta in 1996 each reduced the ability of NGOs and other activists to organize openly in factories. At the same time periodic increases in the government minimum wage, if poorly enforced, combined with the demand for factory jobs, reduced the likelihood that workers would forge the bonds across workplaces, industries, and islands that would have produced a genuinely popular movement. Leadership was also a problem. The criticism that the labor movement lacked "colonels" was all too true. As a consequence it was not the workers who became the much-hoped-for engine for democratic change, and by mid-1998, as Indonesia entered into a period of sharp economic decline and political uncertainty, it was not clear what their role in post-Soeharto Indonesia would be.

Notes

1 See, for example, Daniel Lev, "Human Rights NGOs in Indonesia and Malaysia," in Claude E. Welch Jr and Virginia A. Leary (eds), *Asian Perspectives on Human Rights* (New York: Westview Press, 1990).
2 Hal Hill, "The Economy," in Hal Hill (ed.), *Indonesia's New Order* (Sydney: Allen & Unwin, 1994) pp. 61–3.
3 Tens of thousands remained in detention without charge or trial more than a decade later.
4 Hill, "The Economy," p. 71.

5 Heri Akhmadi, "White Book of the 1978 Students' Struggle," *Indonesia*, 25 (April 1978), 161–4.
6 Adam Schwarz, *A Nation in Waiting* (Sydney: Allen & Unwin, 1994), p. 94.
7 Hill, "The Economy," p. 70.
8 John Bresnan, *Managing Indonesia* (New York: Columbia University Press, 1993), p. 262.
9 Bambang Ismawan in *The Indonesian NGO Agenda*, CESDA-LP3ES (Toronto: 1996), p. 202.
10 Chris Manning, "Survey of Recent Developments," *Bulletin of Indonesian Economic Studies*, 32, 2 (August 1996).
11 Government of Indonesia, National Development Information Office, Indonesia Source Book, 1996, p. 103.
12 Rex Mortimer, *Indonesian Communism under Sukarno* (Ithaca, NY: Cornell University Press,1974), p. 149.
13 Indonesian submission to the US Trade Representative, quoted in Human Rights Watch/Asia, *The Limits of Openness* (New York: Human Rights Watch, 1994), p. 41.
14 Diane Wolf, *Factory Daughters* (Berkeley and Los Angeles: University of California Press, 1992), p. 128.
15 Chris Manning, "Structural Change and Industrial Relations during the Soeharto Period: an Approaching Crisis?," *Bulletin of Indonesian Economic Studies*, 29, 2 (August 1993), p. 81.
16 Editors' introduction to "White Book of the 1978 Students' Struggle," *Indonesia*, 25 (April 1978), 151.
17 T. Mulya Lubis, Fauzi Abdullah, *Langit Masih Mendung* (Jakarta: Sinar Harapan, 1980), pp. 121–6.
18 David Harris (ed.), *Prisoners of Progress: A Review of the Current Indonesian Labour Situation* (Leiden: INDOC-FNV-INFID, 1995), p. 36. The 1130 figure appears to include not just strikes but other forms of work stoppages and walkouts.
19 Juni Thamrin, "Development Policy," in *Prisoners of Progress* (note 18), p. 13.
20 Mies Grijns, Ines Smyth, Anita van Verlzen, Sugiah Machfud and Pudjiwati Sayogyo, *Different Women, Different Work* (Aldershot: Avebury Press, 1994), p. 171.
21 Vedi R. Hadiz, "Workers and Working-Class Politics in the 1990s," in Chris Manning and Joan Hardjono (eds), *Indonesia Assessment 1993* (Canberra: Australian National University, 1994), p. 188.
22 Chris Manning (1993), *op. cit.*, p. 79.
23 "Pidato Pelantikan DPC-DPC SBSI se JABOTABEK," 19 July 1992, in *Serikat Buruh Sejahtera Indonesia*, a booklet published by SBSI in Jakarta.
24 Letter of Muchtar Pakpahan to Pharis Harvey of the International Labour Rights Education and Research Fund, 7 January 1993.
25 For a full discussion of the formation of the PRD and PPBI see Human Rights Watch and the Robert F. Kennedy Memorial Center for Human Rights, "Indonesia: Tough International Response Needed to Widening Crackdown," *Human Rights Watch Short Report*, 8, 8C (August 1996).
26 Dita Sari of PPBI had been arrested a few weeks earlier in Surabaya, East Java, after organizing a rally by workers in ten factories in the Surabaya area on 8 July 1996.

2

Economic and Political Reform in the Philippines, 1986–96: the Evolving Role of Human Rights NGOs

Gerard Clarke

Introduction

On 24 and 25 November 1996, 18 heads of state attended the annual Asia-Pacific Economic Cooperation (APEC) summit at Subic Bay, 150 km northwest of Manila. Formerly a United States naval base with a land area of over 7000 hectares, Subic Bay was transformed into an export-processing zone and ship-repair facility following the termination of the US–Philippines Bases Agreement in 1992. By 1996 Subic had become a successful industrial enclave and was chosen as the summit venue to highlight the Philippines' new-found economic dynamism. At the same time, 400 activists and academics from the Philippines and abroad attended a counter-summit in Manila organized by the Manila People's Forum on APEC (MFPA). Human rights issues figured prominently in the backdrop to the counter-summit. José Ramos Horta, a prominent advocate of the independence of East Timor and winner of the 1996 Nobel Peace Prize, was denied a visa to attend the conference. Provoking further outcry among civil rights campaigners, other activists were denied entry to the Philippines at Manila's Ninoy Aquino International Airport. At Plaza Roma in Manila's Intramuros district, meanwhile, a "tent city" supported by human rights and other NGOs was erected to highlight the eviction of an estimated 250 000 slum-dwellers from their homes in a campaign to improve the city's appearance for the APEC summit.[1]

Together, the summit and counter-summit served to highlight the role of human rights NGOs in the growing debate in the Philippines about human rights and the impact of political and economic reform, especially in the context of the Ramos government's "Philippines 2000" plan. Through an examination of the evolving work of human rights NGOs in the Philippines from 1986 to 1996, this chapter therefore examines the relationship between economic and political reform in the Philippines and the impact on state and civil society.[2] In contrast to other parts of the developing world, the Philippines has experienced a significant process of mutually reinforcing economic and political reform since the collapse of the dictatorship of Ferdinand Marcos in February 1986. By 1996 the Philippines had become a stable electoral democracy characterized by economic growth and falling levels of poverty. This chapter will argue that behind the ostensibly impressive macroeconomic and macropolitical gains, however, acute social, economic and political problems remained, many of them highlighted in the work of human rights NGOs. That is, within the context of economic reforms, human rights NGOs today are dealing with a new form of human rights abuses, that of "development aggression," as they struggle to politically influence the Philippines' economic policy.

Economic and political reform in the Philippines, 1986–96

Since 1986 the Philippines has experienced a profound process of political reform. In February 1986, 14 years of corrupt and authoritarian rule by Ferdinand Marcos ended and Corazon Aquino became President after rigged presidential elections led to a popular revolt. The collapse of the Marcos dictatorship followed the assassination of opposition leader Benigno Aquino in August 1983 and an economic crisis that forced the government to secure structural adjustment loans from the World Bank between 1980 and 1988. The crisis was triggered by deteriorating prices for Philippine exports, especially primary products, from the late 1970s, by conditions of global recession following the 1983 oil crisis, and by the deterioration in international confidence in the Marcos regime following the Aquino assassination. In May 1992 Fidel Ramos was elected as Aquino's successor and assumed the presidency in the first peaceful democratic transition since 1965. Other reforms aided the democratic transition: the elaborate system of "crony capitalism" centered on monopolies and state-owned enterprises was largely disbanded; the military was gradually brought under civilian control; a radical decentralization program was implemented from 1991; and the Ramos government negotiated peace accords with insurgent groups, including the Cordillera People's Liberation Army (CPLA) in 1987 and the Moro National Liberation Front (MNLF) in 1996. By 1996 the Philippines had the most democratic system of rule in southeast Asia, measured by the criteria of a stable competitive party system, regular elections, freedom of association and a free press.

Political reform was accompanied by an equally profound process of economic reform. Until the early 1990s the Philippines was an under-performing, poverty-racked economy, the "sick man of Asia," surrounded by the high-performing tigers and tiger cubs of east and southeast Asia. Like other countries in Southeast Asia, economic growth in the Philippines during the mid-1980s was undermined by the global recession that cut the demand for the region's exports and increased the cost of imported fuel. In addition, however, the Philippine economy was undermined by the graft and corruption of the Marcos regime, poor export growth due to the market distortions of "crony capitalism," and low levels of inward foreign investment. From 1993, however, when a sustained post-stabilization recovery began, gross national product (GNP) grew by over 5 percent per annum, exports by more than 10 percent per annum and real investment by more than 6 percent.[3] According to government figures, two million jobs were created between 1993 and 1995 and the percentage of households living below the poverty line fell from 39.9 percent to 35.7 percent during the same period.[4] By 1996 the Philippine economy was growing at a steady rate and the Ramos government was on target to achieve the objectives of its 1993 "Philippines 2000" plan; Newly Industrialising Economy status by 2000 with per-capita GNP of US$2000 and the percentage of the population living below the poverty line reduced from 50 percent to 30 percent.[5]

In part, economic growth was achieved through the dismantling of cartels and monopolies. Between 1992 and 1995 the Ramos government privatized more than 400 assets and agencies, netting more than US$1.9 billion in 1994 alone.[6] By June 1995 the number of government-owned or controlled corporations had fallen to 101.[7] In other respects, however, attempts to dismantle the old economic order were undermined. Nationalist economic provisions in the 1987 constitution led the Supreme Court to overthrow a number of government measures to liberalize and deregulate the economy while the traditional landed elite inserted loop-holes in tax reform bills. By 1996 anti-dynasty bills proposed by the government were still being obstructed by Congress, especially the House of Representatives. In addition, economic growth was promoted by the growing capacity of the state to generate tax revenue for productive investment and social services.

As these problems suggested, the Aquino and Ramos governments, far from overthrowing the system of political and social rule which developed during the 14 years of the Marcos dictatorship, in many respects restored it. Pre-Marcos "Cacique Democracy,"[8] competitive multi-party democracy based on an elaborate patronage system that enabled a relatively cohesive agrarian-based elite to permeate and to effectively control key political institutions, survived the reform process between 1986 and 1996. In the House of Representatives elected in 1987, 164 of 200 members belonged to the traditional agrarian-elite-based political families.[9] By 1992 the number of representatives belonging to such families had fallen to 145, indicating a small weakening of their position but underlining their enduring power and influence.[10] Similarly, in 1986, when the Aquino government came to power, big landed capitalist families controlled 40 of the 120 leading manufacturing firms in the Philippines.[11] By 1996 the process of liberalization and deregulation had reduced the figure somewhat, but such families retained significant economic, and consequently political, power.

Thus, by the mid-1990s, Philippine democracy was relatively stable and the economy buoyant, but the transition, especially in the late 1980s, preyed on the interests of vulnerable groups long marginalized from political participation and economic opportunity. First, the economic crisis that peaked between 1985 and 1986 pushed up foreign borrowing. In 1988, foreign debt reached US$28 billion, equivalent to US$450 for every Filipino.[12] Debt service, equivalent to 43.9 percent of government expenditure in 1989, pushed down government expenditure on health care, education and social welfare, adversely affecting the rural and urban poor.[13] Secondly, significant land inequality, a traditional source of political instability and market distortion, persisted through the 1986–96 period. In the late 1980s the Philippines had one of the most inequitable land ownership structures in the developing world: roughly 5 percent of families owned 83 percent of all farmland in the late 1980s while 85 percent of the 10 million people directly employed in agriculture were landless in the sense of having no secure title.[14] In recognition, the Aquino government embarked on an ambitious redistributive land reform program from 1987, but the program progressed slowly and behind schedule due to legal loopholes, opposition from a landowner-dominated Congress, and insufficient budgetary resources. Third, in 1987–89, fighting between government forces and the communist-led New People's Army (NPA) resumed, leading to rising human rights abuses nationally and population displacement in affected areas. Since 1989 insurgency has declined substantially but still smolders, leading to on-going displacement.

NGOs were important beneficiaries of this process of political and economic reform but equally they were forced to contend with the adverse social, economic and political consequences of the late 1980s. Between 1984 and 1993 the number of registered NGOs in the Philippines grew by 148 percent to 58 000, more than twice the rate (65 percent) at which registered private sector organisations proliferated.[15] By the early 1990s, NGOs had insinuated themselves deep within the social, economic and political fabric of the country, participating in local government structures, in election campaigns, in contemporary issue-based social movements and in governmental bodies.[16] Within the NGO community, human rights NGOs emerged as the most politically active and militantly vocal. Throughout the years from 1986 to 1996 human rights served as a central issue in a debate about the evolving nature of Philippine democracy and the development model that underpinned it.

Human rights NGOs in the Philippines

In the Philippines, human rights NGOs date to the declaration of martial law in 1972 which enabled Ferdinand Marcos to stay in power beyond the constitutional limit of two four-year terms.[17] After the proclamation of 21 September, Congress was closed, key opposition leaders were arrested, existing political parties were banned, newspapers and radio/television stations were closed or nationalized, and a brutal counter-insurgency campaign was unleashed against rural and urban communities alike. All institutions that existed independently of the state, and that could potentially mobilize opposition to it, were repressed, driven underground or shut down. The Catholic hierarchy and officials of the main Protestant churches supported the regime, yet Marcos clamped down on local social action programs, jailing priests and nuns, raiding church establishments,[18] and disrupting the work of church-based development NGOs.

As a result the Philippine Catholic Church post-Vatican II and a small group of religious activists informed by liberation theology provided the impetus needed to launch the country's first human rights NGOs. Following the establishment of the Church–Military Liaison Committee in November 1973, Citizens' Councils for Justice and Peace were established by church groups in prominent cities and towns, including Cebu and Davao. In 1974 two of the Philippines' largest human rights NGOs were established, Task Force Detainees of the Philippines (TFDP) and the Free Legal Assistance Group (FLAG) and by

1980 a nascent human rights NGO community existed. The human rights movement strengthened further following the 1983 assassination of Benigno Aquino.

Throughout the late 1970s and early 1980s human rights NGOs played a significant role in weakening the Marcos dictatorship: by crystallizing church concern for human rights; articulating a central cause around which the fragmented opposition could unite after 1983, and around which moderate opposition groups could work with the NDF; mobilizing international opposition to the Marcos regime; and building a nationwide human rights movement through grassroots organizations. The Philippines lacks a strong judicial tradition and, as former Supreme Court Justice Abraham Sarmiento notes, the pressure politics of TFDP and other human rights NGOs proved a more effective guarantor of justice under the Marcos regime than the subordinated judiciary.[19] The collapse of the Marcos regime, however, created a dramatic new "democratic space" for human rights NGOs and POs: freedom to organize, liaise openly with other organizations at the local or national level, and actively mobilize new constituencies. As a result, new organizations and coalitions were established while others expanded. In 1987, for instance, the Moro Human Rights Center (MHRC) was established to document the plight of the Moros of Muslim Mindanao. In December 1985 Structural Alternative Legal Assistance for Grassroots (SALAG) was formed by a group of young Ateneo Law School graduates to promote the new concept of alternative law. In 1986 SALAG opened offices in General Santos City (Mindanao) and Baguio (Northern Luzon), widening its reach beyond Manila and among its target clientele of peasant farmers, fishermen, and indigenous peoples, and in 1990 it became a founding member of the new-established coalition of Alternative Law Groups (ALG).

By the early 1990s the Philippine NGO community was extremely heterogeneous in organizational, ideological and regional terms. At the apex stood a small number of NGOs devoted to human rights issues and with a strong presence both in and beyond Manila, including TFDP, FLAG, and the Protestant Lawyers League of the Philippines (PLLP). In 1993, for instance, TFDP had 288 staff and 82 offices in 69 of the Philippines' 72 provinces, making it the largest human rights NGO in the developing world.[20] The next tier was composed of national church organizations and cause-oriented groups with human rights sections or desks. Church organizations included the National Secretariat for Social Action, the National Organization of Churches in the Philippines and the United Church of Christ in the Philippines while cause-oriented groups included the *Bagong Alyansang Makabayan* (BAYAN, New Patriotic Alliance), *Kilusang Mayo Uno* (KMU, the May First Movement), *Kilusang Magbubukid ng Pilipinas* (KMP, Philippine Peasants Movement) and GABRIELA (the General Assembly Binding Women for Reform, Integrity, Equality, Leadership and Action). Regional human rights NGOs, cause-oriented and church groups and coalitions represented a third tier.

Ideologically the human rights NGOs community reflected cleavages in a broad spectrum of social movements. Most human rights NGOs, based in Manila or beyond, most notably TFDP, were identified with the national democratic position and they broadly supported the program of the communist-led National Democratic Front (NDF). A small number of organizations, including FLAG, eschewed identification with the main political tendencies while other NGOs were identified with social democratic or independent socialist parties and groups.

A major development in the period from February 1986 to the early 1990s was the proliferation of regional human rights NGOs and alliances as the political space available to local-level social movements broadened. A national coalition of human rights NGOs, the Philippine Alliance of Human Rights Advocates (PAHRA) was formed in 1986 and provincial and regional branches of PAHRA were gradually established. In addition, organizations and coalitions were formed around strictly local or regional concerns, for instance, the plight of Mindanao's Muslims, the rights of indigenous communities or the destruction of local environments.

The human rights NGO community was also characterized by great disparities in terms of size or income. A relatively small number of human rights NGOs had a genuine national reach or a sizeable or professionally trained staff. Most organizations had little or no paid staff, and relied on volunteers, while cause-oriented groups maintained small human rights "desks" with one or two staff. The prevalence of small organizations or desks made the human rights NGO community highly fragmented and made coalition politics extremely fluid.

Human rights NGOs and the Aquino regime, 1986–92

The collapse of the Marcos dictatorship in February 1986 dramatically transformed relations between the government and human rights NGOs. Within days of assuming office, Aquino ordered the release of most political prisoners, appointed human rights lawyer Silvestre Bello as Justice Secretary and appointed three FLAG lawyers to cabinet-level positions.[21] Highlighting the government's new commitment to human rights, Aquino established a Presidential Human Rights Commission (PHRC) and appointed FLAG Chairman Jose Diokno as Chair and TFDP head Sr Mariani Dimaranin as a Commissioner. Aquino also established a Judicial Reorganisation Committee and a Presidential Committee on Political Detainees, both with TFDP and FLAG representation.

Despite these appointments, however, the popular interests represented by human rights and other NGOs were relatively marginalized by other powerful forces within the Aquino regime. The defection of defence minister Juan Ponce Enrile and Philippine Constabulary head, General Fidel Ramos, in the aftermath of the 22 February elections, precipitated the collapse of the Marcos dictatorship, and both secured senior positions in the new Aquino government, Enrile as Defense Minister and Ramos as Armed Forces Chief of Staff. Allied to the ongoing insurgency, this provided the armed forces with a powerful influence over the new regime, an influence later asserted in five coup attempts between July 1986 and December 1989. In addition, anti-Marcos business interests, many of whom played an active role in the opposition movement that followed the 1983 murder of Benigno Aquino, achieved an important influence as the new government sought to build international confidence and secure donor support. In alliance with international voices in business, government and multilateral circles, these business interests pressured the Aquino government to institute market reforms that deregulated the economy and encouraged exports and inward foreign investment.

Between 1986 and late 1987 relations between human rights NGOs and the government were relatively positive.[22] The armed forces of the Marcos era were reorganised, armed conflict in the countryside abated as insurgent groups assessed the new government, and peace talks were imminent. By 1988, however, peace talks between the government and the National Democratic Front (NDF) had collapsed, the armed strength of the New People's Army was at a peak,[23] and the armed forces increased their leverage within the Aquino government, securing government commitment to a reinvigorated counter-insurgency policy. Social activists, especially NGO workers and human rights lawyers, became significant targets because of their high visibility, especially in campaigns for local elections held in January 1988, and because of a crackdown against NGOs and cause-oriented groups perceived as sympathetic to the triumvirate of the National Democratic Front–Communist Party of the Philippines–New People's Army.

Nationally and in parts of the country where it was the only significant cause-oriented NGO, TFDP was singled out. To intimidate staff at its headquarters in Manila, for instance, the bodies of two dead men were left outside its offices in April 1988. Repression continued throughout 1989, and according to Amnesty International, "dozens of activists involved in lawful non-government organizations which military authorities and some government officials described as fronts for the CPP/NPA disappeared," a significant portion of the estimated 200 victims of extra-judicial killings during the year.[24] Overall, during the two years from late 1987 to early 1989, human rights NGOs and cause-oriented groups experienced repression of an intensity unknown under the Marcos regime, with four FLAG and three PLLP lawyers killed.[25]

Throughout the late 1980s, therefore, human rights NGOs adopted a critical stance *vis-à-vis* the state. TFDP was the most militant and in 1986 it helped establish (PAHRA), widening the flanks of the Philippine human rights movement. TFDP's large staff, nationwide reach, foreign funding and long experience of research and documentation meant that it became an important and credible source of information on the human rights situation, consulted by journalists, politicians and diplomatic officials. Locally, TFDP organized regional coalitions and supported grassroots peoples organizations (POs) such as Families of Victims of Involuntary Disappearances (FIND), Mothers and Relatives against Tyranny and Repression (MARTYR), and SELDA (*Samahan ng mga Ex-Detainees Laban sa Detensiyon at para sa Amnestiya*, the Association of Ex-Detainees against Detention and for Amnesty) mobilizing effective opposition against the government that embarrassed it nationally and internationally. In addition to human rights concerns, TFDP campaigned around a broad range of socioeconomic and political issues including foreign debt, overseas workers and agrarian reform, helping to sustain the united front of the Communist Party of the Philippines. The MHRC, like TFDP a member of PAHRA, played a similar role, documenting the harassment of Moro communities in Mindanao affected by counter-insurgency operations and Muslim migrants in Manila affected by demolitions and slum clearances.[26]

In addition, many human rights NGOs administered socioeconomic programs, many of them supported by foreign donors, to help rehabilitate released political prisoners in the harsh economic climate of the late 1980s or to help political prisoners detained in poor physical conditions.

By the end of 1988, however, as its international image deteriorated, the government attempted to reestablish its democratic credentials. In response, Aquino sought to reestablish links with the human rights

NGOs which had provided some of the harshest and most effective opposition. NGOs endured a tense relationship with the Commission on Human Rights (CHR), established by constitutional mandate in 1987 in succession to the Presidential Commission on Human Rights (PCHR) and the CHR had a negligible impact on the human rights situation in the country. In response, Aquino met with FIND representatives in December 1988 and announced the establishment of a new Presidential Human Rights Committee (PHRC).

According to Aquino, the PHRC was established after "non-governmental organizations, many of them critical of my government's efforts in human rights protection and promotion ... helped government appreciate the gravity of our human rights situation."[27] Headed by the Justice Secretary, and under the control of the Office of the President, the PHRC included representatives from the CHR, and the Senate and House human rights committees. Two seats were allocated to PAHRA and FLAG as representatives of the human rights NGO community. NGOs were initially skeptical of the new body but within a year established an effective rapport. FIND lobbied Aquino for action on behalf of the disappeared in their December 1988 meeting. In response, the PCHR "facilitated searches in camps and military stockades and [brought] cases of ... disappearances to authorities through members of the committee."[28] In 1992, among other measures, the PHRC investigated military counter-insurgency operations in Northern Luzon, and in response to TFDP demands, facilitated the release of political detainees on humanitarian grounds, reviewed the cases of detainees charged with criminal offenses, approved the clustering of political detainees at Manila's Muntinlupa state penitentiary and established a P20 million witness protection program.[29] By facilitating direct relationships between human rights NGOs and officials from the armed forces, justice department and National Bureau of Investigation (NBI), the PHRC avoided the bureaucracy and inertia of the CHR, leading to speedy investigations, direct hearings with human rights victims, and more vigorous prosecutions.

Human rights NGOs and the Ramos regime, 1992–96[30]

The work of human rights NGOs evolved considerably during the years of the Ramos presidency, reflecting the changing character of the Philippine polity and economy and its impact on the human rights NGO community. The years of the Aquino regime heralded a difficult transition from authoritarian rule to electoral democracy and from perennial economic crisis to stable and sustained economic growth. Vulnerable communities, however, bore the brunt of the transition process. Rural peasants suffered especially. Those who lived in areas affected by counter-insurgency operations lost land through land conversions as landlords sought to avoid the land reform program. Alternatively, those who were dependent on government services were affected by budgetary cuts induced by structural adjustment.[31] By 1992, however, the communist-led insurgency had waned considerably under the combined impact of the government's counter-insurgency policy and doctrinal struggles within the Communist Party heralded by the collapse of the former Soviet Union and the Eastern Bloc. This, combined with on-going improvements in the disciplinary record of the armed forces, meant that cases of individual human rights fell significantly. In addition, many human rights organizations cut their socioeconomic programs due to cuts in foreign funding as the economy improved and as donor attention shifted elsewhere. In 1989, for instance, TFDP decided to terminate its income-generating projects, transferring responsibilities to other organizations and closed its center for the rehabilitation of released political detainees. In these circumstances the role of human rights NGOs in the Philippines came under question, but such NGOs found significant new roles as the negative impacts of the new development model became evident.

In reality, the improvement in the human rights situation began long before Ramos came to power. Cases of individual human rights abuse began to fall significantly from the beginning of 1991. But under Ramos, further measures were taken to consolidate the improving rights situation and to cement links with the human rights NGO community. In 1992 Ramos repealed Republic Act 1700, the anti-subversion act, one of the main weapons used under previous regimes to suppress dissent, and legalized the Communist Party of the Philippines. In October of the same year, Ramos established a Fact-Finding Committee on Involuntary Missing Persons to trace 1128 people who disappeared during the Marcos and Aquino regimes and on 3 December 1992, during a meeting with TFDP and FIND, Ramos announced a $4 million compensation fund. In a further step, in a speech to mark international human rights day on 10 December 1992, Ramos acknowledged the legitimacy of human rights NGOs:

> Tension ... between the government and ... human rights [NGOs] is a necessary element in the democracy we are trying to build. We in government must learn to realize that they are not the enemy ... I

> [want] to see both government and NGOs ... work closely together ... [with] more openness between us [and] more empathy with one another's situation.[32]

Ramos' speech illustrated a significant fact: human rights NGOs, many of them opposed to the government of Corazon Aquino, were now open to constructive engagement with the state, while Fidel Ramos, who as Chief of the Philippine Constabulary until 1986 and later as Chief of Staff of the Armed Forces presided over the systematic repression of human rights NGOs, now sought a *rapprochement*. The convergence pointed to changes in the nature of human rights advocacy in the Philippines, to changes in the human rights NGO community, and in turn to changing relations between NGOs and the state.

From the mid-1970s and until the late 1980s–early 1990s, Philippine human rights NGOs emphasized individual civil and political rights in their advocacy and campaign work. As such they based their mandate primarily on the 1966 United Nations' International Covenant of Civil and Political Rights and Articles 2–21 of the 1948 Universal Declaration of Human Rights, often referred to as the "first generation" of human rights. Human rights NGOs also advocated and campaigned on economic, social and cultural rights, the "second generation" of rights enshrined in Articles 22–27 of the Universal Declaration of Human Rights, and in the 1966 United Nations Covenant on Economic, Social and Cultural Rights and on the evolving, though still ill-defined, "third generation" of collective or people's rights enshrined in Article 28 of the Universal Declaration and the 1986 UN Declaration on the Right to Development. Civil and political rights, however, provided the principal *raison d'être*.

The emphasis on civil and political rights stemmed primarily from the nature of the martial law regime over which Ferdinand Marcos presided from 1972 to 1986 and its systematic abuse of individual rights. This emphasis continued for much of the 1980s under the Aquino regime as abuses of individual rights continued, especially from 1987 to 1989 when the government pursued a strident policy of counter-insurgency. Between 1988 and 1991, for instance, FLAG handled 4150 cases, 60 percent involving political detainees charged with offenses relating to rebellion or subversion.[33] From 1990, however, the number of individual cases of human rights abuses fell, and human rights NGOs sought a new *raison d'être*.

In addition, however, many human rights NGOs and grassroots people's organizations that emerged in the Marcos period were sympathetic to the National Democratic Front and provided logistical support wherever possible, for instance, in securing the release of detained activists or guerillas, in campaigning against counter-insurgency operations in areas where the New People's Army was fighting and in supporting the campaigns of mass organizations controlled by the Communist Party (e.g. the frequent *welgang bayan*, or people's strikes of the pre-1986 period). As such, this partisan commitment to a narrowly defined clientele strengthened the commitment to individual, civil and political rights. By the early 1990s, however, the Communist Party of the Philippines was weakening significantly under the impact of a doctrinal struggle between a hardline faction committed to the orthodox Marxism–Leninism–Mao Zedong philosophy of the party's founder José Maria Sison, and a moderate faction that opposed Sison's leadership. Human rights NGOs, an important component of the united front that the Communist Party built up from the mid-1970s, became embroiled in this doctrinal struggle. From the early 1990s a number of NGOs, notably TFDP and PAHRA, began to distance themselves from the Communist Party while others maintained their commitment. In 1994 a formal rift developed over PAHRA's draft Philippine Declaration of Human and People's Rights. At a Congress in October held to discuss the draft, 28 member organizations led by BAYAN walked out. BAYAN denounced attempts "to transform PAHRA into a politically neutral organization which reeks of collaboration and compromise with the state on the issue of human rights violations" while other organizations condemned the PAHRA draft as "too universal to guide human rights advocacy in the [Philippines]."[34] Breakaway organizations formed their own coalition, *KARAPATAN* (Rights) in opposition to PAHRA, and the crisis worsened as prominent human rights organizations fragmented with militant staff forming breakaway organizations. In late 1994, FIND split into two organizations. More seriously in 1995, TFDP split in two as militant staff in regional offices sided with *KARAPATAN* and headquarters staff in Manila with some support from regional staff remained loyal to PAHRA. The split within PAHRA damaged the unity and credibility of the human rights NGO movement and weakened the opposition that it provided to the Ramos government. Equally, however, it prompted organizations such as TFDP to consolidate their autonomy from the NDF and to develop a mandate rooted in international human rights standards.

Human rights NGOs had long been committed to pursuing economic, social and cultural rights but from the late 1980s and early 1990s these rights were pushed to center stage. As economic growth recovered, human rights NGOs working with the marginal communities adversely affected, began to critique the development model on which the recovery was based, especially the "Philippines 2000" plan launched by the Ramos administration in 1992 and the 1993–98 Medium Term Development Plan which implements it. In 1994, for instance, FLAG launched an Economic, Social and Cultural Rights Program and a Environmental Litigation and Advocacy Program. Along with a Death Penalty Program launched at the same time, these programs provided the main impetus to FLAG's work in the mid-1990s.[35] Similarly, in 1993 TFDP altered its constitution to include economic, social and cultural rights and strengthen its mandate with respect to government development strategy.

The 1993–98 Medium Term Development Plan, according to the government, seeks to promote both global competitiveness and people empowerment.[36] At its core the plan commits the government to promoting export-oriented industrialization, especially in tax-free export-processing zones; to improving energy generating capacity in the wake of a 1992–93 energy generation crisis; to improving infrastructure such as roads, ports and airports and to reducing domestic rice production in favor of higher-value crops.[37] The commitment to popular empowerment, allied with the Philippines' responsibilities as a signatory to the 1966 International Covenant on Economic, Social and Cultural Rights, provides human rights NGOs with a mandate to monitor the implementation of the plan, a role to which a number have adapted with gusto. In response, NGOs have claimed that the plan will promote growth at the expense of employment, will damage the already-depleted environment; will displace tens of thousands of people, mostly subsistence farmers and urban slum-dwellers; will alienate the rights of indigenous minorities and will undermine the food security of vulnerable groups.

A central feature of the 1993–98 five-year plan is the creation or consolidation of Regional Industrial Centers (RICs) to promote manufacturing activity for domestic and international markets. Of these, the most important are CALABARZON on the eastern and southeastern outskirts of Manila (named after the provinces of Cavite, Laguna, Batangas, Rizal and Quezon which it straddles) and the Cagayan de Oro–Iligan corridor in northwest Mindanao. The creation of RICs, however, has been accompanied by significant abuses of human rights.

Human rights NGOs played a significant role in supporting farmers displaced by the RIC programme. Between 1992 and 1994 SALAG handled the largest number of cases on an annual basis since its establishment in 1985, mostly representing displaced farmers, along with urban-poor groups affected by demolitions and indigenous communities affected by illegal logging. In CALABARZON many of SALAG's clients were awarded Certificates of Land Ownership Award (CLOAs) under the land reform program but had not received physical deeds due to red tape and corruption. Significant land reclassification powers, devolved to local governments structures in the 1991 decentralization program, strengthened the power of the agrarian elite and stymied the enforcement of CLOAs. Graft and corruption, low wages for judicial officials, and patronage links between judges and politicians militated against SALAG and other human rights NGOs and in most cases land evictions were upheld by the courts.

TFDP and PAHRA branded human rights abuses increasingly associated with the government's macroeconomic development model as "development aggression." The term provoked disquiet among some of TFDP's donors who felt it echoed the militancy of the National Democratic Front. In reality, however, the term reflected a growing concern within the international human rights NGO community with the instrumentalist use of human rights concerns by governments in the quest for economic growth.[38] TFDP and PAHRA continued to campaign on behalf of victims of civil and political human rights abuse, including the 213 political prisoners detained at the end of 1995, the 31 people who "disappeared between August 1992 and December 1995, and on behalf of the human rights and other NGO workers who continued to be harassed, detained and killed."[39] But equally, at a time when the government was proclaiming its commitment to human rights and to working with NGOs, TFDP and PAHRA noted a new character of human rights abuse:

> The most troubling [TFDP] statistics [now] have to do with human rights violations that are not political in nature. During the latter part of Aquino's term [as Philippine President], TFDP began to monitor demolitions in urban areas. The data gathered by the organization in the past three years [1993–95] has been ... alarming.[40]

From 0 percent of human rights abuses in Manila in 1993, TFDP and PAHRA claimed that human rights abuses stemming from "development aggression" in the city, largely demolitions and evictions, rose to 4 percent in 1994 and to 8 percent in 1995.[41] In the first six months of 1995, they claimed, 7082 homes in Manila were destroyed to make way for new roads or for commercial property development.[42] The work of other human rights NGOs was similarly affected. From a focus on agricultural land conversions from 1992 to 1994, SALAG concentrated on urban evictions and demolitions in Manila in 1995 and 1996, while MHRC supported Muslim migrants doubly affected by demolitions and evictions and intense cultural discrimination in the Philippine capital.

TFDP and PAHRA defined "development aggression" as economic interventions by the state or by private entities (foreign or national) with the support of the state, that endangered the interests of whole communities. The sites of such interventions, TFDP and PAHRA claimed, functioned as "breeding grounds" for attacks on civil and political as well as on economic, social and cultural rights, and became "sites of social unrest."[43] While human rights NGOs were implicitly invoking the controversial "third generation" collective of people's rights, in effect therefore, they based their claims on the more substantive "second generation" economic, social and cultural rights. As such, the *praxis* of Philippine human rights NGOs illustrates the time-lag that exists between the theoretical delineation of human rights and their practical defense in the context of struggle between state and civil society.

Human rights NGOs claimed that they were not opposed to development *per se*, but supported the rights of communities to be "active participants," rather than "passive receptors" in the development process.[44] Amid a government commitment to people empowerment in the 1993–98 five year plan, human rights NGOs therefore participated in a debate with the state about the development process and its impact on human rights and, in so doing, challenged the efforts of the state to extend its hegemony over Philippine civil society. To the Philippine government the "Philippines 2000" strategy and the 1993–98 Medium-Term Development Plan are leading to significant improvements in public welfare, including falling poverty rates, improved housing, greater employment and land redistribution. In turn, macroeconomic policy, the government could claim, was underpinned by significant budgetary allocations to a Social Reform Agenda and a safety net to cushion the effects of trade liberalization. To human rights NGOs, however, Ramos government policy was consolidating an inequitable model of development, where only 10 percent of workers were paid more than the minimum wage, where 61 percent of business establishments failed to comply with labor standards governing minimum wages and work hours, and where barely 2.54 percent of the labor force was covered by collective bargaining agreements.[45]

Conclusion

The process of economic and political reform in the Philippines from 1986 to 1996 and the evolving work of human rights NGOs have been closely interlinked. Throughout the late 1980s the Philippines endured a difficult period of economic stabilization and political reform that led to tense relationships between human rights NGOs and the state. NGOs bemoaned the enduring power of the armed forces over government policy, the resumption of a harsh policy of counter-insurgency and a repression of human rights lawyers and organizations unknown under the Marcos dictatorship. Despite these criticisms, however, human rights NGOs participated in a variety of governmental structures at local and national level as NGOs generally asserted their role in the invigorated civil society heralded by the "People Power" revolt of February 1986.

By the early 1990s, however, cases of civil and political rights abated as the communist-led insurgency waned. As a result the role of human rights NGOs in Philippine civil society changed as the concept of human "rights" evolved from the former emphasis on individuals to a new emphasis on collective entities, and from a traditional emphasis on freedoms to a new emphasis on entitlements. From the early to mid-1990s, human rights NGOs focused increasingly on economic, social and cultural rights and the impact of the government's macroeconomic development strategy, especially on the "Philippines 2000" strategy pursued by the Ramos government from 1992. Increasingly, Philippine human rights organizations such as TFDP and PAHRA documented cases of "development aggression," echoing parallel developments in the work of international human rights NGOs. But in the Philippines the particular context of a decline in the number of abuses of civil and political rights, a government macroeconomic development strategy that affected vulnerable social groups, and an ideological split within a large section of the human rights NGO community all served to influence the change in direction.

The political and economic reform process had a complex effect on the state and civil society. In the immediate wake of the collapse of the Marcos dictatorship, civil society flourished and NGOs, including those concerned with human rights, were major beneficiaries, expanding in numbers and influence. In turn, human rights and other NGOs strengthened civil society, *à la de Tocqueville*, by articulating and representing interests that were inadequately addressed by other institutions such as the media, the legal profession, trade unions or political parties. But as the reform process unfurled through the late 1980s and early 1990s, a more Gramscian form of civil society became apparent in two key respects. First, political cleavages within the human rights NGO community weakened their position *vis-à-vis* the state while the enduring power within the state of the agrarian-based elite who have traditionally dominated Philippine politics significantly influenced the strategy of the government *vis-à-vis* human rights and other NGOs. Secondly, the discourse of human rights became a significant new arena of struggle and debate, contested by human rights NGOs but where the state and its supporters exerted a dominant influence. To the extent that civil society expands as the state makes autonomous social forces increasingly subject to rules and laws that are primarily of its design, the proliferation of human rights and other NGOs strengthened the Philippine state and the social forces that primarily benefited from its rule. Equally, to the extent that the proliferation of human rights and other NGOs correlated with the decline of militant social protest, NGOs contributed to the demobilization of anti-state pressure, again strengthening the state.

Human rights NGOs made their greatest contribution in the era of "high politics," the period before and after the collapse of Marcos dictatorship, where militant commitment to democratization and popular empowerment, and concerted struggle on a broad front, ensured they were their defining characteristics. In the Philippines today, human rights NGOs and the issue-based social movements in which they participate retain a significant place in civil society and enjoy significant influence and leverage *vis-à-vis* the state. The political and economic reform process between 1986 and 1996, however, has eroded their position as they adjust to the era of "low politics" where struggles are discrete, issue-specific and fought along a much narrower front.

On the other hand, however, the Philippine human rights movement remains among the most dynamic and influential in the developing world. Human rights NGOs have acquired an important role in monitoring the impact of government macroeconomic policy and development strategy and in supporting those adversely affected by it. The Philippines is a signatory to the 1966 International Covenant on Economic, Social and Cultural Rights and human rights NGOs are closely monitoring the implementation of the "Philippines 2000" strategy to ensure that it complies with the Philippines' international obligations. In the absence of other institutions with a mandate or capacity to monitor government compliance, human rights NGOs are therefore playing an important role in civil society, holding the state accountable to its citizens and defending the interests of marginal groups affected by state policy.

Notes

1 House floor minority leader Ronaldo Zamora estimated that 250 000 slum dwellers were evicted in the weeks leading up to the APEC summit. *Philippine Daily Inquirer*, 5 November 1996.
2 The chapter focuses on the work of four Philippine human-rights NGOs: Task Force Detainees of the Philippines (TFDP); the Free Legal Assistance Group (FLAG); the Moro Human Rights Centre (MHRC); and Structural Alternative Legal Assistance for Grassroots Inc. (SALAG). It is based on research in the Philippines in 1993 and 1996, including interviews with officials of these and other organizations, documentation and publications produced by each of the four organizations, and other background interviews and documentation. I am grateful to the Economic and Social Research Council (ESRC) and the British Academy Committee for South-East Asian Studies for their financial support of the research.
3 NEDA, *The President's 1995 Socio-Economic Report* (Manila: National Economic and Development Authority, March 1996).
4 Ibid.
5 Cf. Fidel Ramos, "Philippines 2000: Our Development Strategy" (pamphlet), Multi-Sectoral Reform Secretariat (Manila, 1993).
6 *The Economist*, 11 May 1996.
7 NEDA 1996, 72.
8 For a more detailed discussion of "cacique democracy", see Benedict Anderson, "Cacique Democracy in the Philippines: Origins and Dreams," *New Left Review*, **169** (May/June 1988).
9 Eric Guiterrez, *The Ties That Bind: A Guide to Family, Business and Other Interests in the Ninth House of Representatives* (Manila: Philippine Centre for Investigative Journalism and Institute for Popular Democracy, 1994), p. 4.
10 Ibid.
11 Temario Rivera, *Landlords and Capitalists: Class, Family and State in Philippine Manufacturing* (Quezon City: University of the Philippines Press, 1994).
12 Rosalinda Pineda-Ofreno, *The Philippines: Debt and Poverty* (Oxford: Oxfam Publications, 1991), p. 2.
13 Ibid., p. 15.
14 James Putzel, *A Captive Land: The Politics of Agrarian Reform in the Philippines*, Catholic Institute for International Relations (New York: London and Monthly Review Press, 1992), pp. 23 and 27.
15 Gerard Clarke, "Participation of Protest: Non-Governmental Organisations and Philippine Politics," PhD dissertation, School of Oriental and African Studies, University of London, 1995, pp. 87 and 305.
16 Gerard Clarke, "Non-Governmental Organisations and the Philippine State 1986–1993," *South-East Asia Research*, **3**, 1 (March 1995).
17 Marcos was first elected in 1965 and reelected in 1969.

18 Robert L. Youngblood, *Marcos against the Church: Economic Development and Political Repression in the Philippines* (Ithaca, NY: Cornell University Press, 1990), pp. 172–3.
19 Abraham Sarmiento, "Philippine Experience in Enforcing a Bill of Rights: Political Realities and Judicial Perspectives," *Human Rights Forum*, **11**, 1 (1992), 33.
20 Clarke, "Participation of Protest," p. 246.
21 Joker Arroyo as Executive Secretary, Augusto Sanchez as Labour Secretary and Rene Saguisag as Presidential Spokesperson.
22 The seeds of the deteriorating relations were sown, however, from the earliest weeks of 1987. On 22 January 1987, 15 000 farmers marched on the Presidential Palace to protest government back-tracking on redistributive agrarian reform. At Mendiola Bridge, police and soldiers opened fire and 18 demonstrators were killed. On 27 January, rebel soldiers launched their third coup attempt against the government, designed to disrupt the referendum on the draft constitution scheduled for 2 February and to end the government's peace overtures to the National Democratic Front–Communist Party of the Philippines–New People's Army (NDF–CPP–NPA). Although the coup was defeated, and the referendum went ahead as scheduled, the coup attempt achieved its other objective, and in a speech on 18 March 1987 Aquino promised to "unleash the sword of war" against insurgents. Miriam Coronel-Ferrer and Antoinette Raquiza (eds), *Motions for Peace* (Manila: Coalition for Peace, 1993), p. 11.
23 The military estimated NPA strength in early 1989 at 24 000 regular guerillas with 10 000 high-powered rifles, and with control or influence over 8000 of the country's 41 000 barangays (villages) (Gregg R. Jones, *Red Revolution: Inside the Philippine Guerrilla Movement* (Boulder, CO: Westview Press, 1989), p. 8). In all probability, however, these figures were slightly higher in mid-to-late 1988.
24 Amnesty International, "The Philippines: a Summary of Amnesty International's Concerns," *ASA 35/02/90* (London: Amnesty International International Secretariat, May 1990), pp. 3 and 5.
25 Clarke, "Participation of Protest," p. 105; LCHR, *Impunity: Prosecutions of Human Rights Violations in the Philippines* (New York: Lawyers Committee for Human Rights, September 1991).
26 MHRC, "A Brief Profile of the Moro Human Rights Centre" (Manila: Moro Human Rights Center, February 1995).
27 Corazon Aquino, "Transparency and Democracy," speech to the United Nations World Conference on Human Rights, Vienna, 16 June 1993, reproduced in *The Manila Bulletin*, 19 June 1993.
28 TFDP, *Pumipiglas 3: Torment and Struggle After Marcos* (Quezon City: Task Force Detainees of the Philippines, 1993), cited in Clarke, "Participation of Protest," p. 292.
29 Ibid., p. 293.
30 Fidel Ramos was elected to a six-year term of office expiring in May 1998. This chapter examines his presidency only to the end of 1996.
31 Although the Aquino government avoided recourse to World Bank Structural Adjustment Loans (SAL) in the post-1986 period, the policy thrust of structural adjustment was maintained by technocrats involved in

the reforms SAL 1 (1980–5) and SAL 2 (1985–8) who remained committed to them and by business interests in the government committed to market reforms.

32 "Ramos Delivers Speech Marking Human Rights Day," Foreign Broadcast Information Service (East Asia), Washington, 10 December 1991, p. 1.

33 FLAG, *The Free Legal Assistance Group, 1974–1994* (Quezon City: Free Legal Assistance Group, 1994), p. 56.

34 BAYAN Press Statement, 26 October 1994.

35 FLAG, op. cit., p. 85.

36 Republic of Philippines, *Medium Term Philippine Development Plan, 1993–1998*, RoP, 1995.

37 Ibid.

38 See, for instance, Ross Daniels ("Cooperation among NGOs: Opportunities and Challenges," paper presented at the International NGO Forum on Human Rights in China, Manila, 30 August–1 September 1996), who uses the term "trading rights for market access" to capture this phenomenon.

39 See PAHRA, *Dark Stains Spreading on Canvas: A Human Rights Report on the Third Year of the Ramos Government* (Quezon City: Philippine Alliance of Human Rights Advocates, 1996)

40 Ibid., p. 19.

41 Ibid., p. 33.

42 Ibid., p. 19.

43 Ibid., p. 34.

44 Ibid., p. 42.

45 PAHRA, "Dark Stains," pp. 50–1.

3

State–Societal Relations in a Rapidly Growing Economy: the Case of Malaysia, 1970–97

Francis Kok-Wah Loh

Introduction

Malaysia has experienced impressive economic growth over the past 25 years. During the first part of this growth, from 1970 to 1985, the state played a crucial role in the economy. Following a short but severe recession in the mid-1980s the state's involvement in the economy was curbed. A series of policies deregulating and privatizing the economy were introduced instead. Liberalization of the economy promoted the necessary adjustments and resulted in unprecedented rates of growth through the late 1980s and 1990s.

The economic turnaround in the late 1980s, and the steady growth over the past 25 years more generally, facilitated "poverty eradication irrespective of race" and "restructuring to abolish the identification of race with economic functions," the twin objectives of the New Economic Policy (NEP, 1971–90). Introduced by Prime Minister Hussein, the NEP was the all-encompassing affirmative action policy which was conducted on behalf of the bumiputeras ("sons of the soil," that is, Malays and other indigenous peoples). Hence important changes to the structure of Malaysia's multi-ethnic society occurred. A Malay commercial and industrial bourgeoisie was spawned and it began to compete as well as cooperate with the entrenched non-Malay, predominantly Chinese, bourgeoisie in the economy decisively. A multi-ethnic middle class also emerged in Malaysia. The ethnic division of labor inherited from colonialism – wherein Malays largely participated in the agricultural sector and lagged behind the immigrant non-Malays who were largely involved in the non-agricultural sectors – was therefore drastically overhauled.

I wish to argue that a culture of "developmentalism" – which valorizes consumerist habits dependent on rapid growth, in turn dependent on political stability guaranteed by the strong state – has emerged in Malaysia which is undergoing late consumerist-industrialization. It is this new political culture especially evident among the new business and middle classes that must be considered. Hence in assessing the lack of, and limited demand for, participatory politics and democratic reforms in Malaysia in the 1990s, a period of rapid economic growth and market reforms, the presence of the strong state is acknowledged, but the importance of a culture of "developmentalism" is stressed, especially among the new business and middle classes. However, I shall contrast the political passivity and the consolidation of the state in the 1990s to the ferment and fragmentation of the 1980s too, to stress that politics in Malaysia was only recently very alive and competitive and that this occurred during a period of recession and initial adjustments to market reforms. In other words, the success of market reforms in the 1990s has not led to political liberalization but toward its demise.

I first discuss the multi-ethnic character of Malaysian politics and the involvement of the state in the economy under the NEP. Second, I discuss the recession and the market reforms beginning from the mid-1980s, subsequently the contrast between the ferment of the 1980s and the withdrawal of the 1990s. Finally, I develop my culturalist explanation.

Multi-ethnic politics and the NEP

About 20 million people inhabited Malaysia in 1997. Officially the people are categorized into bumiputera or indigenous people and non-bumiputera or non-indigenous people. Malays and the aboriginal groups in peninsular Malaysia and the "natives" of Sabah and Sarawak states (both in northern Borneo) totalling about 60 percent of the population, constitute the bumiputera. Chinese (about 28 percent) and Indians (8 percent), who began emigrating to Malaysia during British colonial rule in the nineteenth and early twentieth centuries, comprise the non-bumiputera. Apart from linguistic and cultural differences, religion further distinguishes the ethnic groups from one another. The Malays are invariably Muslims, the Chinese predominantly Buddhists and Taoists, and the Indians largely Hindus with smaller numbers of Muslims and Sikhs. A small minority of Malaysians drawn from all the ethnic groups (including the Kadazan and Dayak bumiputera of Sabah and Sarawak respectively, but excluding the indigenous Malays) are Christians.

Between independence in 1957 until 1969, the Federation of Malaysia was ruled by the Alliance Party, a multi-ethnic coalition government comprising the United Malays National Organization (UMNO), the dominant partner; the Malaysian Chinese Association (MCA) and the Malaysian Indian Congress (MIC). Nominally, it practiced a Westminster system of government and held federal and state elections regularly. For some observers,[1] Malaysia appeared an example of consociational democracy in the developing countries.

In fact, and in spite of regular elections and parliamentary rule, a strong state dominated by the executive had been put into place since independence. However, its control of civil society was achieved not through the use of brute force, the suspension of the Constitution or elections, but through use of coercive legislation passed by parliament. The most important consideration in this regard was how the British expanded and consolidated the coercive powers of the Malaysian (then colonial) state quickly as a result of the communist uprising, euphemistically termed the Emergency (1948–60).[2] It was only with the advent of independence that a set of participatory governmental institutions was introduced, from above. The state's coercive powers granted under the Emergency Regulations were then amended and incorporated into the Independence Constitution and/or other laws for their use in ordinary times in post-independence Malaysia. Ironically, the Malaysian nationalists worked hand-in-glove with the British to expand and consolidate the colonial state, ensuring the defeat of the communist insurrection.

It was upon such strong foundations that the state was further reformed by the post-independence executive to serve its own political ends. Through promulgation of new laws, and amendments to existing ones, this control and domination has been uninterrupted.[3] Consequently, there have developed very few independent organizations capable of advocating the rights and interests of civil society in sustained fashion. Hence the Alliance ruling coalition (and its successor the Barisan) has been able to maintain power in Malaysia since 1957.

However, the absence of consensus outside the Alliance leadership and an economic slowdown in the late 1960s contributed to worsening ethnic relations, culminating in the worst-ever incident of communal riots in May 1969. The riots led to the suspension of parliamentary rule. The National Operations Council, largely drawn from the ruling Malay elites in UMNO and the upper echelons of the administrative and security forces also dominated by Malays, ruled for almost 21 months. Several important changes geared toward promoting Malay interests were introduced before parliamentary rule was restored in early 1971.

First, the Alliance multi-ethnic ruling coalition was expanded to include the erstwhile opposition parties and renamed the Barisan Nasional (BN, National Front). Secondly, various moves to promote a common national identity were undertaken. This included the introduction of a National Cultural Policy (which emphasized Islam and the cultural attributes of the majority and indigenous Malays as the essential bases of the Cultural Policy), the belated implementation of the National Language (Malay) and National Educational policies (including the progressive use of Malay as the sole medium of instruction in secondary schools and universities beginning from 1971) which had been passed in the 1960s, and the pronouncement of the Rukunegara, the national ideology based on a set of five universal principles.

Thirdly, and most importantly for this study, the New Economic Policy (NEP, 1971–90) was formulated and launched. The NEP's premises were that Malay poverty and resentment of inter-ethnic income and wealth inequalities, and the ethnic division of labor which favored the non-bumiputeras and discriminated against the rural bumiputeras, lay at the heart of the May 1969 communal riots. Hence the NEP's "Outline Prospective Plan, 1971–90" proposed that the poverty rate (especially evident among rural Malays) be reduced in peninsular Malaysia from 49 percent in 1970 to 16 percent in 1990. It further proposed that the bumiputera share of corporate equity be increased from 2.5 percent in 1970 to 30 percent in 1990. (With a stake of 60.7 percent in the ownership of assets in the corporate sector in 1970, foreign domination of the economy was clear but local Chinese control was a significant 22.5 percent.) A massive affirmative action program favoring bumiputeras was launched. However, to ensure that non-bumiputera Chinese and Indians were not adversely affected, the NEP was set within the context of achieving rapid economic growth during the duration of the Plan.

There were three aspects to this program. First, allocations for public sector expenditure vastly increased during the NEP years. Whereas only RM4.6 billion was allocated under the pre-NEP First Malaysia Plan (1966–70), RM10.3 billion was allocated under the Second Malaysia Plan (1971–75), the first of the NEP five-year plans.[4] The allocation was increased to RM31.3 billion for the Third Malaysia Plan (1976–80), and to RM48.9 billion for the Fourth Malaysia Plan (1981–85). Under the Fifth Malaysia Plan (1986–90) another RM74 billion was originally allocated but, due to a recession and the introduction of liberal economic reforms, was subsequently revised downwards to RM57.5 billion.[5] Consequently, certain aspects of the NEP were held in abeyance during the late 1980s.

Secondly, with the launch of the NEP in 1971 the state ventured into business, resulting in a proliferation of public enterprises. These were basically of two types: the so-called "non-financial public enterprises" (NFPEs)[6] and the bumiputera trust agencies. The NFPEs, the larger group, were primarily responsible for accelerating socioeconomic development and participated directly in commerce and industry: Petroliam Nasional or Petronas (the national oil corporation), Perbadanan Nasional or Pernas (National Corporation), Heavy Industries Corporation of Malaysia (HICOM), the Urban Development Authority (UDA), the Food Industries of Malaysia (FIMA), and the State Economic Development Corporations (SEDCs) were among the most prominent NFPEs. The trust agencies, on the other hand, purchased and held shares of corporate equity on behalf of the bumiputera community. The major agency performing this task was the Permodalan Nasional Bhd or PNB (National Equity Corporation) and its wholly owned subsidiary, Amanah Saham Nasional (ASN), the national unit trust scheme. Apart from these, there continued the "traditional" state enterprises which provided services such as telecommunications, public utilities and transportation. Taking these together, the number of state-owned enterprises grew from 22 in 1960, to 109 in 1970, to 656 in 1980, to 1014 in 1985, and to 1149 in 1992.[7]

Thirdly, as a result of the introduction of the Industrial Coordination Act (ICA), 1975, and the establishment of the Foreign Investment Committee, the state was enabled to regulate businesses, especially local Chinese enterprises and foreign companies, to ensure compliance of NEP quotas (30 percent reservation for bumiputeras) in a determined manner. Toward this end the minister of trade and industry was given discretionary powers over licensing, ownership structure, ethnic employment, production distribution quotas, local content, and the pricing of products.[8,9]

Public sector employment was further expanded at all levels of government. At the apex could be found federal agencies involved in planning of the NEP, and at the national, state and local levels could be found implementing and monitoring bodies including the public enterprises. In 1970, 11.9 percent (or 398 000 of 3.3 million) of the total employed were found in the public sector. The figures peaked in 1981, when 15 percent (or 757 000 of 5.031 million) of the total employed constituted public sector employees. With privatization and the withdrawal of the state from the economy since the late 1980s, the percentage declined to 12.5 percent (854 000 of 6.84 million) of total employed in 1991.[10]

Inevitably, the various groups of Malaysians differed in their opinions with regard to the NEP. Malays, bumiputera more generally, who benefited from it largely welcomed the restructuring as necessary for fostering national unity while non-Malays tended to resent it as ethnic discrimination, arguing that it worsened ethnic relations. Non-Malay leaders of the Barisan ruling coalition who claimed to support the NEP, nonetheless, were apt to complain about its overzealous implementation by Malay bureaucrats. There was also the important criticism by non-Malays as well as opposition Malay parties that a political patronage system had been spawned as a result of the NEP.[11] At the root of the problem was the availability of business licenses, government projects and soft loans specifically for bumiputeras in accordance with NEP quota requirements. The proliferation of the public enterprises and the setting up of companies by these enterprises further resulted in the creation of numerous company directorships, managerial and executive positions as well as ordinary job opportunities. The most lucrative business opportunities and top positions were often made available to top UMNO leaders, who either acted as trustees on behalf of UMNO or in their own private capacities. In turn, these UMNO leaders made available the less lucrative opportunities and positions to their own supporters. Consequently, UMNO ended up in business. This involvement of UMNO and subsequently other Barisan political parties in business has led to the so-called phenomenon of "money politics" in Malaysia. Indeed, public debates over the NEP were sometimes acrimonious. Yet, in the end, there was general acceptance among Malays and non-Malays of the NEP and for some form of state intervention in the economy on behalf of Malay interests.[12]

Such consensus was facilitated by the steady economic growth that occurred throughout the 1970s. Malaysia's average real GDP growth was 7.6 percent per annum during 1970–80, slowing down to 5.1 percent per annum during 1981–85.[13] Fortuitously, the first decade of the NEP coincided with Malaysia's discovery of offshore petroleum and an international commodities boom. Petroleum exports grew from RM203 million in 1970, to RM861 million in 1975, to RM6709 million in 1980, and to a peak of RM9683 million in 1985. Thereafter, oil prices dropped, bringing petroleum export earnings down to the 1980 value. The export earnings of non-petroleum commodities rose from

RM4.96 billion in 1970, to RM8.37 billion in 1975, and RM21.46 billion in 1980.[14] Such favorable export earnings enabled the government to increase its public, especially foreign, loans[15] which, together with the export earnings, were used to implement the NEP.

But the growth was also spurred on by increased foreign direct investments in Malaysia's export-oriented industrialization program which began in the early 1970s. Free-trade zones were created and various kinds of incentives were offered to attract the foreign companies. Consequently, the structure of the economy changed. Whereas manufacturing accounted for 13.4 percent of GDP in 1970 it was 20.0 percent in 1980 (and maintained at 19.6 percent in 1985, during the worst year of the recession).[16] As a result of the growth of the labor-intensive textile and garments trade, and semiconductor assembly plants in particular, manufacturing employment as a percentage of total employment increased: from 11.1 percent in 1970 to 15.7 percent in 1980, holding at 15.3 percent in 1985.[17]

More generally, employment expanded steadily from 3.34 million in 1970, to 4.8 million in 1980, to 5.47 million in 1985. The unemployment rate declined from 7.8 percent in 1970 to 5.7 percent in 1980, to its then lowest ever rate of 5.0 percent in 1982.[18] Per-capita income rose from RM1100 in 1970 to RM3700 in 1980. By any standards these were impressive growth rates, even more so when the need for inter-ethnic restructuring is considered.

Recession and policy adjustments

The impressive growth, however, experienced a hiccup. The price of commodities including that of petroleum dropped. Subsequently, local as well as foreign investments began falling too. In 1982, when the recession first set in, the government responded by counter-cyclical spending. Public sector consumption, investment and employment were increased. A mild recovery occurred but was followed by a more severe recession in 1985 which lasted until 1986.[19] The federal government launched an austerity drive cutting development expenditure severely and freezing jobs and wages. Many mismanaged and uneconomic public enterprises were closed down. Due to these government measures, as well as private sector retrenchments, unemployment shot up from a low of 5.0 percent in 1982 to 7.6 percent in 1985 and to 8.6 percent in 1987.[20] The government then turned its attention toward stimulating growth.

Significantly, guidelines for foreign equity ownership in manufacturing were liberalized in July 1985. The ICA was amended in December

1985 to make it easier for industrialists to invest in new projects or to expand or diversify their existing investments. In May 1986 the Promotion of Investments Act provided additional tax incentives for manufacturing, agriculture and tourism. By the end of 1986 even more liberal conditions for foreign equity and expatriate staffing were announced in an effort to attract new foreign investments. Another amendment to the ICA followed: this time only firms having more than 75 (previously 25) workers or RM2.5 (previously RM1) million paid-up capital had to comply with NEP requirements. The local Chinese entrepreneurs who had curtailed investments due to the NEP restrictions, and now the recession as well, were being courted to stimulate the economy.

The corollary to deregulation was privatization, which the government hoped would further contribute toward growth.[21] Privatization, broadly defined as "the transfer of government services and enterprises to the private sector" in fact amounted to "denationalization." It involved a wide variety of mechanisms: sale or divestment of public enterprises, public issue of minority or majority of shares in a state-owned public enterprise, sale or lease of public assets, joint public–private sector ventures, private financing of infrastructure projects, allowing private competition where the public sector previously enjoyed a monopoly, etc. Moreover, by awarding privatization projects to bumiputera individuals and businesses, corporate equity ownership could then be transferred to bumiputera interests in accordance with NEP restructuring guidelines. Otherwise, the NEP would be held in complete abeyance, and that would not have been politically expedient.

Privatization began with new projects such as the construction of the north–south toll highway and highway interchanges, and the first commercial television station, TV3. Older, relatively successful public enterprises such as the Malaysian Airlines System (MAS) and the Malaysian International Shipping Corporation (MISC) were privatized next. However, before privatization could get into full swing the worst of the recession was over. The economic growth rate in 1987 was only 5.2 percent but over the next three years, 1988–90, the annual average was over 9 percent. The decline in private sector investments was reversed. Proposed capital investments of all industrial projects in 1990 totaled RM28.17 billion, up from RM5.69 billion in 1985. Foreign investments in industrial projects amounted to RM17.63 billion, up from only RM959 million in 1985. Much of the latter was due to the inflow of investments from Japan, Taiwan, Singapore and Hong Kong due to tightening labor markets, rising production costs at home, and in the cases of Taiwan, Singapore and Hong Kong their loss of GSP status (Generalized System of Preferences) for their manufactured exports to America in particular.[22]

The middle-classes and political ferment in the 1980s

It was during this period of economic recession and initial adjustments to liberal economic policies following a 15-year period of sustained economic growth that new challenges to the state's domination of Malaysian civil society occurred. This new challenge was led by the middle classes. The political ferment was first evident in the form of challenges by the NGOs (non-governmental organizations).[23] Ultimately, the ferment spilled over into the ranks of the trade unions, the Bar Council and the judiciary, and even the political parties, especially UMNO which was politically fractured. It was also evident in the new political vocabulary of "government accountability" and "transparency," of "human rights" and "participatory democracy" that the NGOs promoted.

The structure of Malaysian society had changed as a result of rapid growth and industrialization during the 1970s and early 1980s. The rise of the multi-ethnic middle classes was perhaps most significant. Saravanamuttu, for one, has estimated the size of the middle class. By adding together the total numbers of people involved in professional and technical work, administration and management, sales and clerical work and half of the total numbers engaged in services-related occupations, he estimated that the middle class would comprise just under 36 percent of the total gainfully employed in 1988, up from about 31 percent in 1980.[24] Although one might dispute how Saravanamuttu defines the middle class (for me, middle classes) and his estimate of its size, there is no disputing that a multi-ethnic middle class of substantial size had emerged.[25]

The NGOs

The emergence of the middle classes influenced state–societal relations in Malaysia. Observing political developments in the 1980s several researchers drew attention to the emergence of NGOs which began to usher in a new era of "participatory democracy" and popular demands for government "accountability" and "transparency". Although some of these NGOs had been formed in the 1970s, they had previously operated within limited spheres and avoided public attention.

Conceivably because of increased pressures due to the economic recession and liberalization policy adjustments, the NGOs drew on popular support and came centre-stage in the 1980s. Some major activities of these NGOs are highlighted below.

The first major NGO protest occurred when the government proposed amendments to the Societies Act in early 1981. By classifying societies as either "friendly" or "political," the government sought to prevent organizations, unless they declared themselves as "political," from even issuing public statements, let alone engaging in lobbying or advocacy. The proposed controls to be enforced on the "political societies" included curbs on membership and leaders, and on relations with and funding from foreign organizations. The amendments would also grant the registrar of societies increased powers, and should the minister of home affairs who was proposing the amendments decide to cancel the registration of any society which violated the rules, or reject the application of any would-be society, his decision was "final," i.e. the matter could not be referred to the courts. Incensed by the government's proposed amendments, a group of 19 NGOs and other societies met in Kuala Lumpur on 19 March 1981. Those present included the Consumers Association of Penang (CAP), the Environmental Protection Society Malaysia (EPSM), the Selangor Graduates Society (SGS), Aliran, a Penang-based social reform organization, the Islamic Youth Movement Malaysia (ABIM), several other Islamic organizations, and academic and educational associations. As a result of the meeting, a secretariat run by ABIM, and headed by its leader Anwar Ibrahim (currently deputy prime minister of Malaysia), was established. Although the amendments were passed in parliament (where the Barisan ruling coalition commanded the two-thirds majority necessary for constitutional amendments) in April 1981, the protests continued. In a meeting held the following month more than 700 representatives of 115 societies were present and unanimously resolved to continue efforts to repeal the amendments. By November 1981 over 70 000 signatures against the amendments, the largest-ever signature response for an issue transcending ethnic/religious boundaries, had been collected and presented to the government. There occurred much cooperation across ethnic lines within the secretariat and among the 100-odd organizations. Support for the organizations also came from the Bar Council, the opposition parties, the Democratic Action Party (DAP) and the Islamic Party of Malaysia (PAS), and a major daily, *The Star*, which reported favorably on the protest activities. Finally, in March 1983 the government tabled a second round of amendments to the

Societies Act. Several of the 1981 amendments were maintained, but the particularly controversial one requiring societies to declare whether they were "friendly" or "political" was withdrawn.[26]

Coinciding with the NGOs concerns over these amendments to the Societies Act was a similar anxiety among a smaller group of NGOs, over proposed amendments to the constitution in 1982 which apparently sought to enhance the powers of the executive in its dealings with the monarchy, but which some NGOs interpreted as actually contributing toward further concentration of power in the hands of the Executive. There was occasion for the NGOs to discuss the issues with each other, although no common protest activity was mounted.

Instead, some of the same organizations focused attention on the government's handling of several financial scandals which had resulted in huge losses of public funds as in the cases involving Bumiputera Malaysia Finance (BMF) and the Employees Provident Fund, and of the savings of ordinary Chinese Malaysians who had invested in so-called "deposit-taking cooperatives" associated with MCA leaders. The NGOs pressured the government to set up a Royal Commission of Enquiry so as to get to the bottom of the BMF case. The upshot was the formation of the "BMF Investigative Committee" whose terms of reference and scope of investigation were more restrictive. Due to pressure by the NGOs and the opposition, the BMF Committee's findings were made public ultimately, and belatedly the government removed some BMF officers and cooperated with the government of Hong Kong (where the BMF scandal broke out), to bring others to court.

Another series of criticisms and protests by NGOs concerned environmental issues, the most significant of which centered around the setting-up of the factory, Asian Rare Earth (ARE) Sdn Bhd, which was partly owned by Mitsubishi Chemicals. Set up in 1982 to extract rare trace elements and yttrium from tin tailings, the factory, however, also produced thorium hydroxide, a radioactive waste product. Protests began in November 1983 when the villagers of Papan discovered that the dump for the waste product was to be located just outside their village. Together with villagers from Lahat and Pusing, two nearby settlements, they formed the Perak Anti-Radioactivity Committee (PARC). Their protests were given much media publicity. A group of 16 NGOs including EPSM, SGS, Aliran and the Malaysian Social Science Association who called themselves the Papan Support Group, expressed support for PARC. When the dump was moved to Bukit Merah, a nearby settlement, popular protests again occurred. Subsequently, ARE Sdn Bhd was taken to court by a few residents, and an injunction suspending production, pending the introduction of proper safety measures, was obtained in 1985. The legal expertise for the villagers was provided by a CAP team of lawyers. Popular protests erupted again in February 1987 when the Malaysian Atomic Energy Licensing Board issued a new license to ARE Sdn Bhd to restart production. One particular demonstration attracted support from several thousand people living in the general vicinity of the factory.[27]

The NGOs and other groups worked together in another major campaign when the government introduced, and quickly pushed through, amendments to the Official Secrets Act in late 1986. Under the new law, journalists or others who revealed materials obtained from government sources were liable to stiff fines and/or a jail sentence of up to two years. It was clear that the government was attempting to plug loopholes in the law which had allowed the opposition and government critics to expose government mismanagement and financial scandals. The NGOs organized fora in several major towns wherein speakers questioned the government's intentions and called for the replacement of the OSA with a "Freedom of Information Act." The speakers as well as the audiences in these fora were multi-ethnic. On this occasion the NGOs worked closely with the National Union of Journalists which staged a protest demonstration attracting more than 1000 journalists and together, collected 36 000 signatures protesting the amendments. The protest movement received a fillip when two former prime ministers and the former auditor-general who had headed the BMF Investigative Committee questioned the necessity of the amendments. DAP and PAS, the opposition parties, actively supported this campaign too.

Moreover, there was an effort initiated by Aliran to review the Malaysian constitution in view of its numerous past amendments, each time resulting in further curbs to civil liberties. Apart from the NGOs, Aliran's conference held in August 1987 received warm support from members of the Bar Council, former members of the Bench, and even Tunku Abdul Rahman, Malaysia's first prime minister who declared open the two-day conference. Indeed, he endorsed the call at the end of the conference for establishing a Royal Commission to review the constitution. Later that same month a second conference was organized by the Faculty of Law, University of Malaya. On that occasion the former head of the judiciary, in his opening address, likewise supported the call for a review of the constitution and a return to the spirit of the original 1957 Constitution when a greater semblance of checks and balances, and guarantees of civil liberties, were still in place.[28]

However, the momentum toward participatory democracy and human rights, greater accountability and transparency of government was halted in its stride on 27 October 1987 when the government launched its "Operation Lallang." In this security operation 106 persons – representatives of the NGOs, PARC and Papan Support Group activists, unionists and opposition leaders – were detained under the Internal Security Act (ISA), the draconian law which allows for detention without trial. The publishing licenses of two dailies and two other weeklies were also revoked. In one swift swoop the democratic space that had been slowly claimed by the NGOs was usurped by the state. Over the next year amendments to the Publications and Printing Presses Act and the Police Act were further introduced. The former required printers and publishers to apply for fresh licenses annually, whereas they were only required to apply for renewals previously. With new amendments to the Police Act, police permits were required for holding any political gathering. Individuals caught attending "illegal" meetings were liable to fines and up to one year's jail term.

From NGOs to the Bar and Bench to the political parties

Political ferment and cultural fragmentation initiated by the NGOs was, moreover, manifested in three developments: the consolidation of the executive's power over the judiciary in 1988; the challenge of the Barisan federal government by the opposition-led Sabah state government which demanded greater political and financial autonomy from Kuala Lumpur, and, the political splintering of UMNO itself.

In 1988, shortly after Operation Lallang, the government initiated an assault on the formerly independent, though conservative, judiciary. Due to the contentious nature of several government policies and projects, affected parties had taken the government to court, and on several occasions the judiciary had decided in favor of the plaintiff. Prime Minister Mahathir bin Mohamed himself was particularly irritated by what he considered "judicial interference". On its part, the judiciary had claimed that politics had irrupted in the rule of law. When finally the Lord President (LP, the head of the Supreme Court) sent a letter to the King, expressing concern over what the judiciary considered to be "contempt of court" by the executive, a chain of events was set off resulting in the Lord President's dismissal and the suspension of five other Supreme Court judges who questioned the dismissal of their superior. In an extraordinary meeting of the Bar Council more than 1000 lawyers unanimously passed several resolutions condemning the removal of the Lord President and the five judges, and called for the resignation of the acting Lord President who had sat in judgement of his former superior. Apart from the re-appointment of three of the suspended judges (two of whom were due for retirement within a year) no further redress was forthcoming. Instead, a new team of Supreme Court judges was appointed. The powers of the judiciary were further curtailed by the introduction of new "ouster clauses" which denied judicial review of the executive's decision and interpretation of laws such as the ISA, the Broadcasting Act, the Publications and Printing Presses Act and the Societies Act.[29]

In the second case, pressure was brought to bear upon the executive in the federal Barisan government by the Parti Bersatu Sabah (PBS), a multi-ethnic party dominated by Kadazans (mainly non-Muslim and the largest bumiputera community in Sabah).[30] The PBS' first electoral contest was in 1985. It challenged and successfully defeated the Barisan-Berjaya state government led by Harris Salleh which had been in power since 1976. Launched less than a month before the election, the PBS possessed little funds and had no political organization to speak of. But it attracted a core of well-educated middle-class Kadazans who expressed much anxiety about their own and their Sabah's future. These anxieties became the electoral issues the PBS sponsored: Harris's increasingly authoritarian rule, his pro-Malay Muslim cultural programs which the Kadazans viewed as discriminatory, and the question of state rights. PBS leaders condemned Harris for ceding the island of Labuan to the federal government without prior consultation with the Sabah people.

In the following state election, held only a year later, the question of state rights was again highlighted. But equally pertinent were the destabilizing activities of Harris and another former Chief Minister linked to the Barisan federal government, especially their attempt to by-pass the electoral process and seize power illegally, and the bombings, illegal demonstrations and rioting (attributed to the followers of these two leaders) that occurred just prior to the 1986 elections. Without backing from the federal government the PBS marched home to victory.

During the rest of the 1980s the PBS continued to harp on the issue of state rights, and attempted to retrieve a greater measure of political autonomy and fiscal independence for Sabah, more federal development allocations for Sabah in spite of the recession on account of its slower pace of growth, etc. Just prior to another round of state elections in July 1990, several Sabahans linked to PBS leaders were detained under the ISA. According to the federal government the individuals had been involved in secessionist activities. The PBS denied the allegations. Significantly, state rights, yet again, emerged as the principal issue in the election. The federal Barisan leaders denounced the PBS for "fanning parochial sentiments," "playing with fire," and being "anti-federal." The PBS responded that the emotive theme of "Sabah for Sabahans" was consistent with the original constitutional guarantees accorded Sabah at the time of Malaysia's formation in 1963 but which had been eroded by previous pro-Barisan state governments over the years. In the event, PBS secured yet another comfortable victory in the July 1990 state elections, suggesting popular support in the struggle for state rights.

Finally, there occurred a challenge to the executive which developed from within UMNO, the dominant ruling party. Compared to the NGOs', the judiciary's and the Sabah government's challenges, this was surely the most serious of all. Tengku Razaleigh, formerly the Minister of Finance, challenged Prime Minister Mahathir for the UMNO presidency. Earlier, Mahathir's deputy, Musa Hitam, had stepped down from his Cabinet position as Deputy Premier due to reportedly "unresolvable differences" with Mahathir, but promptly threw his support behind Razaleigh.

As party elections approached, in April 1987, the party split down the center with equal numbers of ministers allied to the contenders. Mahathir, however, managed to seize a slim victory and proceeded to purge the cabinet of all Razaleigh supporters. In retaliation, some of Razaleigh's allies claimed irregularities in the party elections and took the matter to court. The courts declared UMNO an "illegal organization," forcing Mahathir to form "UMNO Baru" (or New UMNO). Razaleigh's defected faction formed a new Malay-based party, Semangat 46. For the first time in Malaysia's political history an alternative Malay party with widespread support stood in opposition to the ruling party. Several by-elections resulted in the victory of candidates affiliated to Razaleigh, indicating support in several constituencies, and facilitating the formation of a multi-ethnic opposition front with serious hopes of replacing the Barisan ruling coalition as the 1990 general elections approached.

Many of the NGOs and the major trade unions came out with a joint manifesto which all but endorsed the opposition front called "Gagasan Rakyat." In an election tainted with blatant manipulation of ethnic sentiments by the Barisan ruling coalition,[31] and in what has been termed by a team of observers from the Commonwealth as "free but not fair" elections, the ruling coalition regained its two-thirds majority in parliament (though it only polled 53.4 percent of the votes in Malaysia's "first past the post" electoral system). Having curbed the NGOs, re-asserted the executive's prerogative over the judiciary, and regained control of UMNO, the ruling regime now restored its domination of parliament as well.

Importantly, political ferment and cultural fragmentation occurred in the 1980s at a time of economic recession and state withdrawal from the economy. Privatization and other policies of economic liberalization were in their initial stages, the consequences of which had not yet been felt by most. The leadership of the groups involved in the various challenges of the executive was clearly middle class and had been spawned by the steady economic growth that had occurred in Malaysia between 1970 and the mid-1980s. The patronage system that developed as a result of the state's involvement in the economy, in turn promoted by the launching of the NEP, had led to nepotism and mismanagement of public enterprises. The recession of the mid-1980s was instrumental in bringing to light many of the financial scandals which had previously gone unnoticed. Retrenchment and the job and wage freezes that occurred further threatened livelihoods. As development allocations in the Fifth Malaysia Plan were revised downwards, some groups in Sabah and Sarawak, already claiming discrimination in the allocation of federal funds against their states during ordinary times, now felt an even greater sense of discrimination. Though the economic liberalization policies adopted were welcomed by the non-Malays, since they opened up new economic opportunities for them, the same policies further aggravated the anxieties of Malays (as well as Kadazans in Sabah and Dayaks in Sarawak), who had been beneficiaries to government largesse and subsidies under the auspices of the NEP, throughout the 1970s. With a shrinking public sector, the NEP virtually being held in abeyance, and initial privatization projects benefiting a small group of Malay entrepreneurs, it was no wonder that Razaleigh, though ultimately defeated, was nonetheless able to muster widespread support from almost one-half of UMNO's leadership as well as rank-and-file.

Renewed growth and privatization in the 1990s

Despite the political challenges, the National Development Policy (NDP, 1990–2000), which replaced the NEP had two objectives: poverty eradication and restructuring within the context of economic growth. The strategy for economic recovery introduced in the 1980s would be continued in the 1990s, with the private sector replacing the state as the engine of economic growth and privatization accelerated.

Indeed, by 1995, 374 privatization projects had been implemented. According to the government these projects had saved it RM72.7 billion in capital expenditure and removed almost 97 000 workers from the government payroll.[32] Privatization projects under the Seventh Malaysia Plan (1996–2000) included the Bakun Hydroelectric Power project, the money-losing Perwaja Steel Bhd, Bank Bumiputera (the largest bank in the country), and several coastal highway projects, ports, and light rail transit systems in other towns. Additionally, corporatization (a variation of the privatization theme which allows for administrative autonomy without sale of government assets) of the major hospitals, the state universities, the Public Works Department, several technical institutions and tourism-related bodies had also begun.

Fed up with the waste, inefficiency and corruption associated with the public sector, a large section of the Malaysian public, especially the middle classes, had initially welcomed the policy shift. Anticipating opportunities for themselves, businessmen – bumiputera, local Chinese as well as foreigners – also welcomed privatization.

However, the distribution of the benefits of privatization has been uneven. Firstly, the most lucrative privatization projects appear to have been awarded to a small group of bumiputera and non-bumiputera individuals and companies possessing the right personal and political connections.[33] This unhealthy development occurred because the open tender system was largely replaced by the so-called "negotiated tender system." Controversial award decisions have been explained in terms of a "first come, first served" policy, i.e. in favor of those who first identify "viable" privatization proposals to the government.[34]

Consequently, privatization has led to the deepening of the political patronage system, initially associated with the allocation of licenses, government projects and soft loans to bumiputeras as required under the NEP guidelines. It now also includes favored non-bumiputera businessmen. Since many of them are related to the ruling elites and their political parties, it has further entrenched the involvement of Barisan parties and politicians in business.[35]

The second criticism of the privatization policy has been increased charges for various services now provided by the private sector, but without improvement of the services provided. In part this is because privatization has often replaced public monopolies with private monopolies.

The "IWK fiasco" is a case in point.[36] In 1993, Indah Water Konsortium (IWK) was awarded the entire national sewerage system after signing a 28-year concession agreement costing RM6 billion with the government. Beginning in 1995 many households, industries and commercial concerns began to be charged new rates, the basis of which was unclear to the consumers. Compared to the rates previously charged (which were often included as part of annual property assessment rates, or as in the state of Penang, collected monthly by the Penang Water Authority on the basis of the number of toilets found on the premises) the new IWK rates, apparently based on the amount of water consumed, were exorbitant. Regardless of the time IWK started billing consumers, the new rates applied from the time IWK took over control of the sewerage treatment system from the local authority which previously ran the service. Hence the new bills included arrears: this was so even when services were not yet rendered in certain locales. IWK's argument was that its charges were based on "accumulated services" which consumers might not yet enjoy but "would do so in time." As IWK assumed control of more sewerage treatment systems, and began billing more consumers, more complaints were voiced. Responding to the public outcry the Federation of Malaysian Consumers Association (FOMCA) launched a campaign to collect one million signatures of protest. Within six months after the launch some 500 000 signatures had been collected. The opposition parties and other NGOs joined in the protest. Finally, in May 1996, after several Barisan elected representatives and state governments voiced their concern, the deputy premier ordered IWK to submit a fresh proposal regarding its charge rates, requiring that the rates be commensurate with the quality of services actually being provided to consumers.

At any rate, privatization and other market reforms facilitated rapid growth and furthered structural changes to the economy. The manufacturing sector grew: from 27 percent of GDP in 1990 to 33 percent in 1995. Manufacturing employment as a percentage of total employment has also increased from 19.5 percent in 1990 to 25 percent in 1995. The growth of the manufacturing sector can be further seen in the changing composition of Malaysia's exports. In 1970 manufactured goods accounted for only 12 percent of gross exports from Malaysia. The corresponding figures for 1980 and 1990 are 22 percent and 59 percent, respectively. However, the figure jumped dramatically to 80 percent for 1995. Electronic and electrical products accounted for 65.7 percent of total manufactured products while textile and footwear, wood products, chemicals and chemical products accounted for another

13.95 percent in 1995. The index of poverty has been further reduced from 17 percent in 1990 to 11.1 percent in 1995, and unemployment is down to its lowest ever in Malaysia's history.[37] Indeed, the last 10 years of uninterrupted economic growth averaging over 8 percent annually has transformed Malaysia into, arguably, East Asia's "fifth tiger," after the Republic of Korea, Taiwan, Hong Kong and Singapore. But it has also made Malaysians, especially the business and middle classes, enamored with not only market-driven development but consumerism as well.

Mass consumerism and the discourse of the individual

Mass consumerism is especially evident among the business and middle classes in urban areas. But the rural folk have not been impervious to this aspect of developmentalism either. This consumerism is influenced by so-called "western" tastes and lifestyles as communicated to Malaysians through advertisements and through the global media and entertainment industry generally.

In this regard the revamping of the major Malaysian newspapers, almost all controlled by the Barisan parties, in the early 1990s was no accident.[38] Segmentation of the newspapers occurred: in addition to the usual local, regional, international and sports mainstay there are nowadays business sections with stock market and company reports (something altogether new for the Malay dailies), and a "Life and Times" or its equivalent essentially catering for middle-class interests.

Much of the same has also happened over television. All the stations report on the stock market and business developments in Malaysia daily. These are supplemented by the transmission of reports of business developments outside the country via arrangements with CNN, etc. on all weekdays. Advertisements are presented throughout the day. Moreover, television shopping programs also broadcast the sale of items which may be ordered via the telephone and paid in instalments by use of the credit card. Hence Malaysian newspapers, and the mass media generally, have become virtual organs of the government on the one hand, and advertising copy on the other.[39]

How has this market-driven growth and consumerism influenced the course of politics in Malaysia in the 1990s? Clearly, the Barisan government had consolidated its rule by the mid-1990s, with the apparent acquiescence of different sectors of Malaysian society. No doubt the use of force and coercive laws by the state facilitated this consolidation.

Perhaps less obvious is the emerging culture of "developmentalism" – the Malaysian variant of which in the 1990s valorizes uninterrupted economic growth driven by the market, widespread consumerism, "discourse of the individual" and political stability.

The politics of developmentalism

Central to my argument of a shift from the political ferment and participatory politics of the 1980s to the consolidation of Barisan rule and the politics of developmentalism in the 1990s is the enhanced role of the consuming individual which has furthered the "discourse of the individual" in Malaysia.[40]

The state, through its individuation (or atomization) of civil society into independent subjects (with personal identity cards, income-tax payments and Employees Provident Fund contributions and savings, and of course the franchise) is partly responsible. But, above all, mass consumerism promotes persons becoming individualistic. For one consumes consciously as an individual, not as groups or as communities.

A second aspect of this discourse of the individual is individuality, by which is meant expression of one's own autonomy, freedom and identity usually related to personal achievement – perhaps in education and in one's career. But success also fosters wealth and allows one to enjoy and sustain the consumerist lifestyle that has become so much part-and-parcel of the business and middle classes' everyday lives. Consumables become, as it were, emblems of their achievement, identity, and even their "freedom." In fact, such expressions of identity are often a function of marketing, telecommunications and advertising in the late twentieth century, and not exclusively one's own, but shared.

For our purpose this notion of freedom is most pertinent; it is, at best, of a very private nature, far removed from the original sense of political liberty. Indeed, the discourse of the individual in Malaysia in the 1990s has not resulted in individualism, that ideological orientation associated with a struggle for the individual's civil and political rights, so crucial for the realization of liberal democracy, which developed in Western societies with industrial capitalism, but prior to the emergence of mass consumerism.

No doubt many Malaysians fear the coercive laws, especially the ISA. Nonetheless, an equally if not more important reason why Malaysians have rallied behind the Barisan government is because they valorize political stability nowadays. For stability is considered essential for uninterrupted economic growth, for rising incomes, for maintaining and fulfilling consumerist lifestyles, and for expressing one's individuality, and apparent "freedom" from a private and personalized point of view. This is particularly the case among the middle and business classes. It is probably only when the social issues directly impinge on the individual's economic and social well-being that some measure of participatory democracy is evident. Otherwise, many controversies in the 1990s – financial scandals and corruption, environmental destruction, resettlement of urban pioneers – have not elicited the same kind of sustained popular support they did in the 1980s.

It is this twist in the discourse of the individual in late modernity in Malaysia, associated with market-driven development and mass consumerism, that explains political developments in the 1990s: support for the Barisan and the stability it provides on the one hand, political withdrawal from controversy on the other.

Conclusion

Initially the state played a crucial role in the economy. However, its role has been severely curbed due to market reforms introduced since the late 1980s. Although recession occurred in the mid-1980s, nonetheless, the average growth rates achieved throughout the past 27–8 years have been adequately impressive and have spawned the consolidation of the middle classes. The middle-classes began to push for more participatory democracy and governmental accountability. Political ferment occurred among the NGOs, the Bar and Bench, the unions, and even political parties during the 1980s. However, the consolidation of Barisan rule and of the executive's power over civil society in the 1990s highlight that the ferment was an aberration, probably due to the recession which brought to light many of the financial scandals and incidences of mismanagement involving politicians and public enterprises. It further caused anxieties because of retrenchment, wage freezes and withdrawal of subsidies, and intensified competition and conflict within elite circles including in UMNO itself.

Hence, contrary to re-emergent modernization theorists, the market-driven reforms and the rapid growth that Malaysia experienced in the 1990s did not lead to political liberalization but to its demise.

Notes

1 For instance, A. Lijphart, *Democracy in Plural Societies* (New Haven, CT: Yale University Press, 1977).
2 See Francis Kok-Wah Loh, *Beyond the Tin Mines: Coolies, Squatters and New Villagers in the Kinta Valley* (Malaysia: Oxford University Press, 1988).
3 See Johan Saravanamuttu *et al.*, "The Pitfalls of NICdom," *Asian Exchange*, 8, [1/2] (1992), 13–48.
4 Prior to August 1997, before the currency crisis afflicted East and South-East Asia, RM2.50 was the equivalent of US$1.00.
5 K. S. Jomo, *Growth and Structural Change in the Malaysian Economy* (London: Macmillan, 1990), pp. 106–7, 111, 162–3.
6 The NFPEs were previously known as off-budget agencies (OBAs). They are so classified because their budgets are not reflected in the budgets of the federal and state govenments.
7 Ozay Mehmet, *Development in Malaysia: Poverty, Wealth and Trusteeship* (London: Croom Helm, 1986), p. 11; Edmund T. Gomez, "Changing Ownership Patterns, Patronage and the NEP," in Muhammad Ikmal Said and Zahid Emby (eds), *Malaysian Critical Perspectives: Essays in Honour of Syed Husin Ali* (Kuala Lampur: Persatuan Sains Sosial Malaysia, 1996), pp. 135–36; and M. Rugayah (1995), "Public Enterprises," K. S. in Jomo (ed.), *Privatizing Malaysia: Rents, Rhetoric, Realities* (Boulder, Co.: Westview Press, 1995), p. 66.
8 James Jesudason, *Ethnicity and the Economy: The State, Chinese Business and Multinationals in Malaysia* (Singapore: Oxford University Press, 1989), pp. 135–40.
9 Predictably, local Chinese entrepreneurs found the ICA irksome, especially since it originally applied to any manufacturing firm with RM100 000 (after the 1978 amendment, RM250 000) or more in shareholder capital or employing more than 25 workers.
10 Rugayah, "Public Enterprises," p. 65; Mehmet, *Development in Malaysia*, pp. 9–10; Gomez, "Changing Ownership Patterns, Patronage and the NEP," p. 136.
11 Ibid.
12 Khoo Khay Jin,"The Grand Vision: Mahathir and Modernisation," in Joel S. Kahn and Francis Kok-Wah Loh (eds), *Fragmented Vision: Culture and Politics in Contemporary Malaysia* (Sydney: Allen & Unwin, 1992), pp. 44–76.
13 Jomo, *Growth and Structural Change*, p. 38.
14 Ibid., pp. 54–5.
15 Ibid., pp. 184–90.
16 Ibid., p. 43.
17 Ibid., p. 79.
18 Ibid., p. 79, and Rugayah, "Public Enterprises," p. 65.
19 For details see K. S. Jomo, Boo Tiek Khoo and Yii Tan Chang, "Vision, Policy and Governance in Malaysia," in Leila Frischtak and Izak Atiyas (eds), *Governance, Leadership and Communication* (Washington, DC: World Bank, 1996), pp. 78–91.
20 Jomo, *Growth and Structural Change*, p. 79.
21 The best source available on the privatization policy and exercise in Malaysia is K. S. Jomo (ed.), *Privatizing Malaysia: Rents, Rhetoric, Realities* (Boulder, Co.: Westview Press, 1995).

22 Jomo, Khoo and Chang, "Visions, Policy and Governance," pp. 84–5.
23 For a brief survey of Malaysian NGOs and comparison with their counterparts in the South-East Asia area, see Francis Kok-Wah Loh, "ASEAN NGOs in the Post-Cold War World," in J. Lele and W. Tetty (eds), *Asia – Who Pays for Growth?* (Aldershot: Dartmouth Publishing, 1996), pp. 41–61.
24 Johan Saravanamuttu, "The State, Ethnicity and the Middle Class Factor: Democratic Change in Malaysia," in K. Rupesinghe (ed.), *Internal Conflict and Governance* (New York: St Martin's Press, 1992), pp. 48–9.
25 Mohd Nor Nawawi, "Dasar Ekonomi Baru, Pepaduan Nasional dan Kelas Menengah," in Khadijah Muhamed and Halimah Awang (eds), *Dasar Ekonomi Baru dan Masa Depannya* (Kuala Lumpur: Persatuan Sains Sosial Malaysia, 1991), pp. 140–54; Halim Salleh *et al.*, "Dasar Ekonomi Baru dan Persepsi Etnik di Kalangan Pemimpin Pertubuhan," in Khadijah Muhamed and Halimah Awang (eds), *Dasar Ekonomi dan Masa Depannya* (Kuala Lumpur: Persatuan Sains Sosial Malaysia, 1991), Jesudason, *Ethnicity and the Economy*; Muhamad Ikmal Said, "Malay Nationalism and National Identity," in Said and Emby (eds), *Malaysian Critical Perspectives*, pp. 34–73.
26 K. S. Gurmit Singh, *Malaysian Societies: Friendly or Political?* (Petaling Jaya: Environmental Society of Malaysia, 1984).
27 Tan Sooi Beng, "The Papan-Bukit Merah Protest," in CARPA, *Tangled Web: Dissent, Deterrence and the 27 October 1987 Crackdown in Malaysia* (Kuala Lumpur: Committee against Repression in the Pacific and Asia, 1988).
28 Aliran, *Reflections on the Malaysian Constitution* (Penang: Aliran Publications, 1987).
29 Lawyer's Committee for Human Rights, *Malaysia: Assault on the Judiciary* (New York: Lawyer's Committee for Human Rights,1990), pp. 20–32.
30 Drawn from Francis Kok-Wah Loh, "A New Sabah and the Spell of Development," *Southeast Asian Research*, **4**, 1 (1996), 63–83.
31 Mustafa K. Anuar, "The Malaysian 1990 General Election: the Role of the BN Mass Media," *Kajian Malaysia*, **8**, 2 (1990), 82–102; and Tan Chee Beng, "Resorting to Ethnic Games (Again)," *Aliran Monthly*, **11**, 1 (1991), 20–4.
32 See the remarks for Dr Gan Khuan Poh, senior director of the Economic Planning Unit of the Prime Minister's Department, *The Star*, 24 May 1996.
33 For instance, the government sold 32 percent of its stake in the Malaysian Airlines System worth RM1.8 billion in 1993 to Tajudin Ramli of Technology Resources Industry. In 1994 the Bakun Hydroelectric Power project was awarded *in toto*, without tender, to Ekran controlled by Ting Peck Khiing.
34 Gomez, "Changing Ownership Patterns, Patronage and the NEP."
35 See Edmund T. Gomez, *UNMOs Corporate Investments* (Kuala Lumpur: Forum, 1990) and his, "Changing Ownership Patterns, Patronage and the NEP."
36 This discussion is based on FOMCA, "In Search of a Million Signatures," *Aliran Monthly*, **16**, 4 (1990).
37 Ministry of Finance, *Quarterly Bulletin of Statistics*, 1996.
38 Francis Kok-Wah Loh and Mustafa K. Anwar, "The Press in Malaysia in the 1990s," in Said and Emby (eds), *Malaysia Critical Perspectives*, pp. 96–131.
39 Ibid.
40 The distinctions made here are drawn from K. Abercrombie, N. S. Hill and B. Turner, *Sovereign Individuals of Capitalism* (London: Allen & Unwin, 1986), p. 2.

Part II
Africa

4

Civil Society, Structural Adjustment and Democratization in Kenya

Julius E. Nyang'oro

Introduction

"Political transition" are the new buzzwords in studies of Africa's political economy. Like all buzzwords in academic discourse, there is some empirical basis for the claim that Africa has undergone and has experienced significant changes in its political landscape since the beginning of the 1990s decade, as will become apparent in my discussion below. One thing is also clear, however: political changes in Africa are part of a global movement toward political liberalization. In Africa most of the authoritarian regimes–military or one-party systems – have liberalized partly in response to popular protests, but also as a result of international forces.

These developments provide a hope that the crude authoritarianism of most states in Africa of the post-colonial era will become a thing of the past, and that a firm foundation for the practice of more liberal (democratic?) politics has now been established for the region. The initial basis for this hope is that even in the most authoritarian states of the past such as Zaire (now called the Democratic Republic of Congo), Togo, Malawi, to mention a few, the nature of the political discourse has fundamentally changed.

It is important to note that, while there are encouraging signs for the liberalization of politics across Africa, there are also significant shortcomings in this process to warrant a pause in our enthusiasm and celebration of "democratization" in Africa. Thus, my analysis of Kenya is meant as a reflection in the significance of these broad regional/continental changes with a special emphasis first on political parties, and second on the role of civil society, both at the formal and informal levels. Even though Kenyan civil society is relatively weak and highly fractured, it has played an important role in the struggle for political liberalization. Due to its weakness, based largely on rising ethnic divisions, formal political parties will continue in the near future to be the most important players in the transition to democracy. How successful civil society in Kenya becomes may be, in the final analysis, the answer to the question of how deep democratic change in Kenya has been. The first step in my analysis is to examine the transition from a single-party to a multi-party political system and the role of civil society in such transition.

Post-colonial Kenya

Kenya received her independence from the British on 12 December 1993. The war of independence, the Mau Mau, was led by Jomo Kenyatta, and resulted in the death of 10 000 Africans and a few dozen whites. After the bitter war ended, Kenyatta became the country's first president and remained so until his death in mid-1978. He was succeeded by the current president, Daniel Arap Moi. Thus, Kenya has only had one change of government since independence.

In 1982 there was an attempted coup against Moi and, from that point onwards, Kenya became a classic one-party authoritarian state. A constitutional amendment, Section 2a, officially made Kenya a *de-jure* one-party state. With the resultant increased political repression after the coup attempt, Kenya's civil society became actively involved in the struggle for the removal of Section 2a and the implementation of greater political freedoms, including elections.

Multi-partyism and the December 1992 elections

In December 1991 Kenya officially announced the repeal of the constitutional provision (Section 2a) which had made the country a one-party state since 1982. The announcement by President Daniel Arap Moi's regime of this change followed the now much publicized suspension of economic aid by Kenya's principal foreign donors – the US, Britain, France, and the World Bank – at their consultative meeting in November 1991 in Paris. The donors gave the Moi regime six months to institute meaningful and substantive political and economic reforms before funds could be released. In the meantime the US government announced a travel advisory to Americans intending to visit Kenya during the peak tourist season in December. The reasons for the travel advisory were that continuing political unrest posed physical dangers to foreigners visiting Kenya after an American tourist was killed by unknown assailants earlier in the year.[1] The combined effect of the aid suspension and the potential loss of revenue from the tourist industry – Kenya's largest earner of foreign exchange – were enough to make the Kenyan government take a second look at its commitment to one-party rule. Some political detainees were released and Mr Moi announced the arrest of Nicholas Biwott and the late Hezekiah Oyugi, respectively former Minister of Energy and former Permanent Secretary in the Office of the President in charge of internal security.

Under normal circumstances the dismissal and subsequent arrest of these two men would not raise an eyebrow; but these were not ordinary men in the political power structure in Kenya. Besides President Moi, Messrs Biwott and Oyugi were perceived to be the two most powerful people in the country. Mr Biwott through his business and ethnic ties to the president was for all intents and purposes the "president's man." Mr Oyugi, on the other hand, was privy to all state secrets, especially as they related to security and the human rights situation which had been a point of contention with aid donors. Both Biwott and Oyugi had been implicated in the disappearance and subsequent murder of former foreign minister Robert Ouko in April 1990.[2] Thus the arrest of Biwott and Oyugi was a major concession on President Moi's part to pressure from foreign donors. However, a few weeks after their arrest, the two men were released for "lack of evidence." Biwott remained in parliament as a backbencher while Oyugi was transferred to a new post as Chairman of General Motors (Kenya). Oyugi died in London on 8 August 1992 of a brain ailment. The cause of his death is still surrounded in controversy, as are all deaths of prominent people in Kenya.[3] Mr Biwott has since been politically rehabilitated and, as of the end of 1997, he held the powerful post of minister in the President's Office.

At first glance it would seem that the decision to allow political pluralism and a multi-party system in Kenya was based on pressure from foreign donors, and indeed newspaper reports have suggested that to be the case.[4] While this argument may have some validity, I would like to suggest that the principal push for political reform was internally generated. I would argue that the seeds of opposition to one-party rule in Kenya were planted ironically by the Moi-led Kenya People's Democratic Union (KADU) before independence in 1963. KADU's *raison d'être* was to voice the concerns of smaller ethnic groups in Kenya which did not seem to have influence within the Kenya African National Union (KANU) – the majority party. The merger of KANU and

KADU in 1964 (essentially an absorption of KADU into KANU) signalled the convergence of petty-bourgeois interests which had found expression in the two parties before independence.[5] Kenya thus became a *de-facto* one-party state in 1964.

From the very beginning, however, it was clear that the newly enlarged KANU was not a happy camp, as Oginga Odinga, leader of the opposition until his death in January 1994 and Vice President under Kenya's first president, Jomo Kenyatta, was later to report in his autobiography, *Not Yet Uhuru* (1967).[6] Ajulu, among others, has characterized the conflict within KANU at that time as between a radical faction of the petty-bourgeoisie (represented by people such as Odinga, and that of a traditional/embryonic bourgeoisie represented by Kenyatta) with regard to the trajectory of Kenya's political and economic development. The formation of a new political party, the Kenya People's Union (KPU), in 1966 under the leadership of Odinga signaled the culmination of the rift between the two factions of the bourgeoisie. KPU was banned in 1969, and Kenya, for all practical purposes, remained a one-party state until December 1991.[7] In line with this general point, Jennifer Widner has brilliantly analyzed the process by which there was a shift in government–party relations when the political system moved from a single-party dominant system in which KANU remained a loosely organized "debating society" with little policy influence under the first Kenyan President Jomo Kenyatta (1963–78) toward a Kenyan "party-state" in which KANU increasingly became a vehicle for the President's Office to control political opposition. This political centralization meant that holding an important position under the President's Office guaranteed tremendous political power. Therefore, as I noted earlier, the position of the late Hezikiah Oyugi, for example, can be explained only by the centralization of power in the President's Office and in turn by the increasing authoritarianism of the state.[8]

The point of emphasis, however, should be that throughout the post-independence period there has been no consensus in Kenya on the desirability of a one-party system. Indeed foreign pressure on the Moi government on the question of human rights has always been in reaction to internal struggles – by students, lawyers, politicians, clergy, etc. Thus the massive riots by various segments of the population which took place in July 1990, primarily in the Central and Rift Valley Provinces in response to the banning of public meetings called to advocate political pluralism, must be seen as reflecting this longstanding tradition of opposition to one-partyism.[9] The call by Smith Hempstone, the American Ambassador to Kenya, in 1990 and in subsequent months for the release of political prisoners was based on evidence provided by, among others, members of the Kenya Law Society, whose leadership in the past few years has included people such as Paul Muite, who until recently was Vice-Chairman of the leading opposition party in Kenya, the Forum for the Restoration of Democracy–Kenya (FORD–Kenya). Paul Muite – along with world-famous archeologist/nature conservationist Dr Richard Leakey – is a founding member of a new political party in Kenya known as SAFINA (The Ark). Due to the Kenyan government's machinations, SAFINA was not formally registered until November 1997, thus ensuring that its presidential candidate would perform poorly in the general elections of 29 December 1997 given SAFINA's inability to organize legally across the country.[10]

Even though the Moi regime took its time to officially register SAFINA as a legal political party, the regime had been compelled by both domestic and international pressure to register seven opposition parties in 1992. The three strongest opposition parties during the 1992 elections and to date are FORD, which split into two factions: FORD–Kenya and FORD–Asili (original); and the Democratic Party (DP). FORD–Kenya was led by Oginga Odinga until his death in January 1994. Odinga came from Nyanza Province where the party has the strongest support. Ethnically, Nyanza province is overwhelmingly Luo, which comprises about 13 percent of the Kenyan population. The current leader of FORD–Kenya is Kijana Wamalwa, a lawyer from the Western Province. Ethnically, the Western Province is overwhelmingly Luhya, which also comprises about 15 percent of the Kenyan population. FORD–Asili is led by Kenneth Matiba, an old politician from the Central Province (Kikuyu land) where the party has its strongest support. DP is led by Mwai Kibaki, a veteran politician from the Kenyatta era. He was at one time Vice President under Daniel Arap Moi. DP's strongest base of support is also the Central Province. The Kikuyus represent about 20 percent of the Kenyan population.[11]

To answer the call for multi-party elections, President Moi declared in early 1992 that general elections (presidential and parliament) would be held in December. The opposition parties were still busy fulfilling registration requirements while KANU and Moi were already gearing themselves for the elections in December. As late as August 1992, FORD–Kenya, for example, did not know who its presidential candidate was going to be. Indeed the split between the two FORD factions was caused by Kenneth Matiba's refusal to accept the number two position under Oginga Odinga. Thus going into the December 1992 elections, even though the opposition had done a tremendous job mobilizing the population against President Moi and KANU, the latter won the elections and were poised to triumph again in the next elections slated for late 1997. The disunity among the opposition, which for the most part is ethnic in basis, has presented a major obstacle for the opposition in its attempt to dislodge President Moi in spite of his unpopularity.[12]

Organizational difficulties and disunity among the opposition parties on the basis of ethnicity aside, it would seem that one of the key issues in the continued political liberalization in Africa generally, and in Kenya in particular, is how well civil society in its fractured state responds to the pressures and results of political liberalization itself. To be sure, we have had less than half a decade of multi-partyism in Kenya, and we may have to wait a few more years before a meaningful assessment of political change can be made. I think it is best to guard against false optimism lest we become disillusioned by what we may see as too slow a change in the transition from authoritarianism.

Civil society and the struggle for democracy in Kenya

The notion of civil society has gained currency in the contemporary discourse on democracy in Africa. In general political analysis, however, civil society as a concept is not particularly new. Patrick Chabal, among others, has suggested that civil society is now being introduced into the political analysis of Africa largely "because the (over)emphasis on the role of the state has led to a definition of the political sphere as being virtually conterminous with the politics of the state."[13] The crisis of the post-colonial state and the popular demand for less centralization has led the analysis of African politics to move away from "the state" as a primary focus to activities within civil society. Indeed, as numerous studies have shown, analyses of politics in Africa of the past decade or so had lagged behind "action on the ground" in the sense that, for a significant portion of the population in African countries, the daily lives of individuals were conducted substantially in the vast area of activity away from the state that is understood to be "civil society." Thus, to most analysts of politics in Africa, the success of democracy on the continent is primarily (if not solely) dependent on how vigorously civil society is able to maintain its independence/autonomy from the state.[14]

Even though civil society as a concept is now widely used in the discourse on politics in Africa, it is still a concept that raises significant analytical and practical problems. A leading African intellectual, Samir Amin, for example, notes that it is difficult to discuss civil society outside the framework of capitalist social relations.[15] For Amin, the room for activity by civil society is predetermined by its function in relation to capital accumulation. Once there is a contradiction between civil society's activities and those of the need for capital accumulation, the former must be adjusted to suit the latter. Amin argues that unless this inherent contradiction between forces of capital and those of the general "masses" who aspire to become "really" free is resolved, there cannot be real democracy. Liberal democracy, according to Amin, is thus limited by its incapacity to effectively challenge capitalist inequalities. Civil society in this context is not and cannot be dynamic enough to be the essential building block for democracy that is popular. Thus if we follow Amin's argument to its logical conclusion, the basing of democratic hopes in Africa on "civil society," as understood in the bourgeois sense, leaves a lot to be desired. Patrick Chabal quotes at length another leading Africanist scholar, J. F. Bayart, who is struggling to understand the concept of civil society in relation to Africa. I reproduce the quote here to emphasize the complexity which careful analysis of both the concept and phenomenon of civil society demands:

> Though it is arguable that the concept of civil society is not applicable outside European history I shall define it provisionally as "society in its relation with the state. ... insofar as it is in confrontation with the state" or, more precisely, as the process by which society seeks to "breach" and counteract the simultaneous "totalization" unleashed by the state. ... The notion of civil society is thus an ambivalent (and not just conflictive), complex and dynamic relation between state and society. [Civil society] is not necessarily a discrete entity completely external to an equally discrete source of power. That much is obvious in the case of institutions or organizations which represent civil society within political society, such as parliaments, parties or trade unions. But it is equally true of the power structures themselves, these are by no means immune to the particularistic pulls of civil society. ... Moreover, civil society is not necessarily embodied in a single, identifiable structure. It is by its very nature plural and covers all sorts of different practices; any unity there is requires human creativity. Finally, civil society is not merely the expression of dominated social groups. It encompasses not only popular modes of political action but also the claims of

> those socially dominant groups which are no less excluded from direct participation in political power.[16]

Of particular interest to me in the above quote is the recognition that civil society is not necessarily embodied in a single, identifiable structure, and that it is not merely the expression of dominated social groups. It is a recognition that civil society as an arena is a nasty place which reflects all the contradictions in society. The interests of capitalist, professional associations, non-governmental organizations, trade groups, etc. are all represented in civil society and thus access to the state by members of civil society may be differentiated relative to the power held by particular groups. This observation points to the danger that I see in the celebration, and/or actual romanticization of civil society by Africanist scholars. The limitation of civil society as the midwife of democracy thus must be acknowledged.[17]

This is not to suggest, however, that civil society, both as a concept and its real activities, are not useful to the understanding of contemporary Africa and the continuing opening up of the political and social space. As the French traveller Alexis de Tocqueville observed in relation to America in the last century, civil society is extremely important in defining the nature and extent of democracy in society. If we understand civil society to be a sphere of social interaction between economy and state, then there is an inverse relationship between the state's capacity to coerce the governed and the extent to which local/particularistic interests organize themselves to advance their objectives. The more the members of society organize themselves into groups to advance their particular interests, the less likely the state can function in an autonomous and unaccountable manner. As Joel Barkan has observed, the proliferation of organized interests is a bulwark against unbridled state power.[18] This autonomy may be one of the key principles in the building of democracy.

Yet, as I have noted, there is an inherent danger in the over-celebration of civil society as the basis for democratic change given the relative strength of various groups in society. It needs to be pointed out that not all states are bad, and neither is the state always the enemy. Historically, the state has played a positive role in political mediation between various groups in society and has also played a progressive role especially in defense of capitalism *vis-à-vis* the reactionary forces of feudalism.[19] Thus, as we examine the role of the state in Africa, we need to specify the particular dynamic in which social forces are operating. The experience and practise of most states in Africa, however, has been that of undermining democracy, thus allowing for the generalizations that we encounter about the positive nature of civil society. At this juncture, therefore, as the transition to more liberal politics in Africa continues, and in spite of all the problems associated with the concept of civil society, activities within civil society are still a good measure of how "open" and "pluralistic" a political system is. Conversely, the "depth" of democratic change in Africa may be measured by the extent of its autonomy from the state.[20]

In the specific case of Kenya, despite its fragmentation, civil society has played an important role in pushing for pluralistic politics and the minimizing of state centrality in people's lives.[21] As Widner has indicated, state centrality was characterized principally by the legal banning of opposition parties and the creation of a "party-state." Political participation was restricted to the ruling party KANU. Given these formal restrictions it is fair to say that most of the opposition to the regime was primarily informal. Participants in the opposition reflected the various interests in society: business and professional associations, the church, students, women's organizations, and number of non-governmental organizations (NGOs).[22]

In Kenya, two groups stand out as having played an active role in opposing the Moi regime before December 1991: the church and students, especially university students. To a lesser extent, various women's organizations assumed an activist role in opposing the Moi regime, but this was after the lifting of the ban on multi-party politics in 1991, and most women's organizations actually concentrated on mobilizing women for political participation in the 1992 elections.[23] Stephen Ndegwa has also argued that NGOs for the most part became active in politics only after their existence was threatened by the regime. Otherwise, many NGOs were content to do their business in essentially an apolitical way. Thus the students' activities – organized under the Association of University Students – and the church will stand out in this regard. Two examples on the role of the church should suffice.[24]

First, following the riots of July 1990 to demand multi-partyism, the Church of the Province of Kenya (CPK) (the Anglican church) took issue with the government in terms of the causes of the riots. The CPK's clergy, especially Manases Kuria, then Archbishop of Nairobi, Henry Okullu, then Bishop of Maseno South in Nyanza Province, and the late Alexander Muge, Bishop of Eldoret in Rift Valley Province, demanded that the government face up to the economic and political needs of the people. Bishop Okullu specifically asked the government to resign to clear the way for a new beginning. Thus, in a Sunday sermon on 15 July 1990, Bishop Okullu asked the government to convene a constitutional assembly and called for a transitional government of national unity. Bishop Okullu justified his position by arguing:

> Who has driven us to this ugly predicament where we were tempted to go back to a state of backwardness, [something] we could never have thought of [ourselves]? The blame must lie on the KANU Government which, in spite of all calls to arrange for a national convention, a dialogue with all the people and a democratic form of government, has insisted on going its own way.[25]

Bishop Okullu concluded his sermon by condemning the detention without trial of former Cabinet Ministers Kenneth Matiba and Charles Rubia, who had been picked up by the police for allegedly inciting people to riot. The two former ministers were not released until early 1992. Mr Matiba in the meantime suffered a stroke while he was in detention and had to be flown to London for emergency medical treatment. As we have seen, he later became the leader of one of the opposition parties, FORD–Asili.

Second, the National Council of Churches of Kenya (NCCK) has been in the forefront over the years in opposition to the KANU politics of exclusion. The NCCK is a cross-denomination umbrella organization of churches. Most recently the Catholic Church has joined forces with the NCCK in making pronouncements against the state. The NCCK's leadership is formed from the member churches through open elections. Originally the NCCK solely dealt with church and social services, but it became more politically conscience- and protest-oriented with the increasingly oppressive nature of the state and the impact of SAPs on the population. The NCCK has both continental and international linkages, and receives additional funding from international church networks, including the World Council of Churches.

Along with its constituent members, NCCK on various occasions called for the abolition of the queuing procedure which had forced people at polling stations to queue behind their preferred candidate's picture or his or her representative, thus violating the principle of the secret ballot. This procedure was "exclusionary" because only those in support of the preferred KANU candidate would venture to the polling stations. Thus exclusionary politics here were both by design and by default. As a result of the adoption of the queuing procedure, voter turnout in the 1988 general election fell precipitously to only 24.6

percent of those eligible, the lowest in Kenya's history.[26] In the period since the allowing of multi-party politics in Kenya, NCCK has produced educational materials urging people to exercise their constitutional right to vote, sometimes with a demonstrated preference for the opposition. As the NCCK further participated in the political process, including the attempt at reconciling the divergent opinions within FORD, its former General Secretary Rev. Samuel Kobia was threatened with bodily harm. The speculation was that these threats had connection to the KANU regime.[27]

NCCK has continued to play a significant role in channeling political demands in the current phase of the transition by organizing a commission to study the causes and effects of the "ethnic" clashes which took place in the first half of 1992, accounting for several thousand deaths and the displacement of between 255 000 and 300 000 people, especially in the Rift Valley Province which is President Moi's stronghold.[28] A parliamentary Commission headed by Mr Kennedy Kiliku, MP for Changamwe (Coast Province), to investigate the clashes actually came to the same conclusion as the NCCK report. Both reports implicated high government officials, especially William ole Ntimama, Minister for Local Government; Ezekiel Barng'etuny, nominated MP; and Nicholas Biwott, MP and former Minister of Energy.[29] Thus the role of the church in the struggle for pluralism and government accountability has been an extremely important one. An interesting development in recent years has been the active role played by the Catholic Church in speaking out against state oppression. Previously this mantle had been taken up by the NCCK and the Anglican Church.[30]

Church clergy have concentrated much of their agitation through sermons, as well as pronouncements by individual clergy outside of the church context. The NCCK has also hosted joint meetings and seminars with other pro-democracy groups, particularly lawyers' organizations. An extremely important aspect of their work has also been political education. Before the 1992 elections the NCCK held seminars on why it was important for people to vote. The organization focused on the national responsibility of every citizen to vote. After the elections the NCCK continued to play a leading role in political education until June 1997 when a Nairobi meeting specifically dealing with education was brutally broken up by the police forces. While the international community was originally outraged, and IMF and World Bank funds to Kenya were temporarily suspended, aid was shortly resumed in order (ironically) to put the "teeth" back into international condemnation of the authoritarian regime's undemocratic practices and corruption.

Similarly, the Association of University Students was under constant threat of being banned. It focused its political activities on demonstration, the boycott of classes, and the issuing of leaflets in support of underground pro-democracy movements such as Mwakenya. The NCCK's success lies partially in the fact that it is officially not an NGO. The state had wanted the NCCK to register as an NGO. The NCCK strategically refused to do so, claiming that it was not an NGO but a church organization. As such, it has not been subjected to the draconian NGO act and has had the ability to stand out as a church organization with the moral authority to criticize the government.

Unlike in neighboring Tanzania, where the regime was willing to entertain multi-partyism before political mobilization by the opposition proceeded too far, the regime in Kenya has resisted multi-partyism at every opportunity. Indeed, as reports by domestic and international human rights organizations have indicated, those fighting for multi-partyism in Kenya have suffered physical abuse at the hands of the regime.[31] I would further argue that this maltreatment of advocates for democracy has actually engendered a "high-strung" passionate culture of opposition, including violence. As an example, some members of the two FORD factions were actually engaged in physical combat during the first elections for national party office-holders of FORD–Kenya. Followers of FORD–Asili had intended to disrupt FORD–Kenya elections until scores of them were arrested by the police for public disturbance.[32]

Another example is the reaction of the followers of the unregistered Islamic Party of Kenya (IPK), who made the coastal city of Mombasa, the stronghold of Islam in Kenya, virtually ungovernable for about a week when the government refused to register the party, allegedly because of its religious base and "exclusionist" tendencies.[33] In mid-1993 there were further disturbances on the Kenyan coast, thus continuing the tradition of violence in Kenyan politics. In mid-1997 more violence erupted. This time it was not quite clear what was the immediate cause of the violence. What was clear, however, was that the target of the violence were people who were perceived to be "foreigners;" that is, people from provinces other than the Coast Province. This violence in many ways resembled that of the Rift Valley Province, which a few years earlier had resulted in several thousand deaths and hundreds of thousands of displacements for people who were seen to be "foreigners" in Rift Valley.[34] I would argue that the culture of intolerance has been dangerously nurtured in Kenyan politics as a result of the long-standing decision by the government to control every avenue of political life. Thus the stone-throwing incidents at the FORD elections, the riots in Mombasa, and the so-called ethnic clashes in the Rift Valley Province are no accidents.

But perhaps the biggest obstacle to state building and democratization in Kenya is the problem associated with ethnicity and the use of ethnicity for political purposes. Much has been said about the colonial policy of divide-and-rule in Africa. The policy was based on favoring one ethnic group over others in terms of providing education or introducing development projects which then would be the basis for one ethnic group supporting the colonial enterprise. This of course had the ultimate effect of weakening the anti-colonial struggle because the nationalists not only had to overcome the obstacles created by the colonial state, but had to also deal with ethnic disunity among Africans.[35] In Kenya the big ethnic divide was between the major and more numerous ethnic groups: Kikuyu and Luo, which formed the basis of the earlier KANU; and the smaller ethnic groups such as the Kalenjin, Maasai, and Luhya. The smaller groups formed the basis of the earlier KADU. As we have seen, the two parties, KANU and KADU, merged under KANU at the time of independence. The fact that the initial politics of the predominant parties was essentially ethnic, the future of KANU, and indeed the reality of the post-1991 period has been heavily influenced by ethnic considerations.

In the aftermath of the new multi-party era in Kenya, all the major political parties, in spite of official denial, have used ethnicity as the primary basis for political mobilization. FORD–Kenya is heavily Luo-biased (although it has national appeal); FORD–Asili is primarily Kikuyu as is the Democratic Party. KANU now reflects the smaller ethnic groups, which in the later years of colonization and the early years of independence had allegiance to KADU. Even though previous analyses of Kenyan politics had attempted to move beyond ethnicity as a basis for political organization, it seems that ethnicity as a mobilizing force is still a reality.[36] Githu Muigai has argued that:

> [in] the absence of other platforms upon which to base political appeal, ethnicity presents itself as the most natural basis of political organization, feeding as it does on pride and prejudice and on the "herd instinct."[37]

Given the realities of ethnic politics in Kenya, and in the rest of sub-Saharan Africa, it seems, that far from political liberalization being a force to unite civil society across ethnic groups, it has actually provided an opportunity for ethnic leaders and politicians to revive and use ethnicity for narrow political ends. Thus the failure of the Kenyan opposition to unite under one leader in the presidential ballot can largely be attributed to ethnic dynamics within DP, FORD–Asili, and FORD–Kenya. Simple arithmetic tells us that the total number of votes for President Moi in 1992 were 1 962 666 while the combined opposition received 3 437 458. Clearly, if the opposition had united under one candidate, Moi would have been defeated. It follows, therefore, that for purposes of state building (and democratization?) ethnicity in Kenya has played a negative role, and has been continuously used by leaders both in and out of government to create divisions among the population. Thus even though civil society, which cuts across ethnic lines in Kenya, is fairly active and has played a significant role in bringing about political change, its efforts are still hampered by the reality of ethnic politics.

Structural adjustment and the Kenyan economy

Like most other African countries, Kenya has been under one form or another of structural adjustment for the past two decades. What is interesting about the Kenyan case is that, in the 1960s and 1970s, Kenya was held up by Western countries as the example of how well capitalism can succeed in Africa. On a constant basis Kenya was compared to neighboring Tanzania, which had made a formal commitment to socialism and "self-reliance." As the 1970s came to a close the Tanzanian economy was almost in total shambles while the Kenyan economy showed a steady growth of about 6 percent a year. Nonetheless, as Colin Leys and others suggested in the mid/late 1970s, Kenya's structural problems were deeply rooted in the international inequalities between the North and the South, and as such was a time-bomb which was bound to explode in terms of:

1. population growth far exceeding economic growth;
2. too much external dependence for industrialization;[38]
3. corruption associated with state centralization;[39] and
4. the crisis in the global economy occasioned by the 1973/74 price increases.[40]

The literature on rent-seeking activities is fairly clear on the detrimental effects it has on economies.[41] To a great extent the Kenyan state has been managed by a rent-seeking class which in the final analysis exacerbated the structural problems in the economy. Naturally, the Kenyan state has been forced to deal with declining fortunes in its economy by instituting economic reforms which are popularly known as structural adjustment programs (SAPs). SAPs were first tried in the 1970s, and then abandoned and reimplemented in 1983. Studies by the World Bank show that the biggest problem facing economic reform in Kenya in the past decade has been what Gurushri Swamy has called the "patchy and intermittent" commitment to the policy. According to Swamy, "the stated policies ... were undermined by implementation that was often lethargic and sometimes contrary to the stated policies."[42]

Of course, as reform packages elsewhere have been, the most important issue regarding SAP in Kenya has been the extent of state intervention in the economy. While Kenya has over the years stated officially that it would adhere to "free-market" principles, state intervention in the agricultural sector, for example – which employs the majority of the population in Kenya – has been constant and thus has led to periodic cancellation of credit facilities by the World Bank.[43] Since the late 1970s the Moi government has also consistently attempted to effect economic redistribution in favor of Moi's ethnic group, the Kalenjin. Coupled with the charge that the Kenyan government printed money to finance the 1992 elections, it seems that the Kenyan economy is increasingly on shaky ground as reform is continuously undermined by vested interests at the state level. If this behavior by the state is translated at the level of civil society, it would seem that the latter has had limited capacity to influence state action, and therefore has suffered the ravages of structural underdevelopment disproportionately.

Conclusion

The political economy of Kenya is at a crossroads. Civil society, while active, is still in its infancy and is highly fractured on the basis of ethnicity. Kenya's NGOs suffer from poor organization and weak/inexperienced leadership. In addition, many of the NGOs are uni-focused on the goal of removing the current regime, without giving due thought to what may come next. The economy has suffered a number of setbacks, primarily due to the state's incapacity to adjust to the twin demands of civil society needs and managing the economy itself. This of course places the democratic project in danger of failing. But there is ample evidence (creation of SAFINA, for example) that people have not given up in their struggle against the authoritarian state. While economic reform is still haphazardly implemented, the open question is whether economic reform (liberalization) can/will undermine efforts to liberalize politics. This is a particularly important question in Kenya because of the ongoing pressure from international financial institutions for the state in Kenya to loosen its grip on the economy. The centrality of the state in resource distribution (rent-seeking), however, is making the implementation of SAP a tall order indeed. Further, fractured civil society almost guarantees that only the formal political parties will continue to be the major players in the politics of transition. The balance sheet that comes out of this complexity is one of limited reform both on the political and economic sides of the equation.

Notes

1 *New York Times*, 30 November 1991.
2 For an account of this controversy see Julius E. Nyang'oro, "The Quest for Pluralist Democracy in Kenya," *TransAfrica Forum*, **7**, 3 (Fall 1990), pp. 73–82.
3 *Society* (Nairobi), 24 August 1992.
4 *New York Times*, 4 December 1991.
5 Rok Ajulu, "Kenya: the Road to Democracy," *Review of African Political Economy*, **53** (March 1992), pp. 79–87.
6 Oginga Odinga, *Not Yet Uhuru: An Autobiography* (London: Heinemann, 1967).
7 See Michael Chege, "Swapping Development Strategies: Kenya and Tanzania after the Founding Fathers," in David E. Apter and Carl G. Rosberg (eds), *Political Development and the New Realism in Sub-Saharan Africa* (Charlottesville, VA: University of Virginia Press, 1994), p. 252; and Ajulu, "Kenya," pp. 79–87.
8 Jennifer A. Widner, *The Rise of a Party-State in Kenya: From "Harambee!" to "Nyayo!"* (Berkeley, CA: University of California Press).
9 Julius E. Nyang'oro, "The Quest for Pluralist Democracy," pp. 73–82; and James N. Karioki, "University of Nairobi and the Demise of Democracy in Kenya," *TransAfrica Forum*, **7**, 3 (Fall 1990), pp. 83–93.
10 *Daily Nation*, Nairobi (various issues 1996–97).
11 Norman Miller and Rodger Yeager, *KENYA: The Quest for Prosperity* (Boulder, CO: Westview Press, 1994), p. 72–74.
12 Githu Muigai, "Ethnicity and the Renewal of Competitive Politics in Kenya," in Harvey Glickman (ed.), *Ethnic Conflict and Democratization in Africa* (Atlanta, GA: African Studies Association, 1995), pp. 161–96.
13 Patrick Chabal, *Power in Africa: An Essay in Political Interpretation* (New York: St Martin's, 1994), p. 82.
14 See, for example, Joel Barkan, "Resurrecting Modernization Theory and the Emergence of Civil Society in Kenya and Nigeria," in David E. Apter and Carl Rosberg (eds), *Political Development;* and Julius O. Ihonvbere, "The 'Irrelevant' State, Ethnicity, and the Quest for Nationhood in Africa," *Ethnic and Racial Studies*, **17**, 1 (January 1994), 42–60.

15 Samir Amin, "Preface: The State and the Question of 'Development'" in Peter Anyang' Nyong'o (ed.), *Popular Struggles for Democracy in Africa* (London: United Nations University Press and Zed, 1987), pp. 1–13.
16 Quoted in Patrick Chabal, *Power in Africa,* p. 83.
17 See contributions in John W. Harbeson *et al.* (eds), *Civil Society and the State in Africa* (Boulder, CO: Lynne Reinner, 1994).
18 Joel Barkan, "Resurrecting Modernization Theory," p 111.
19 Ellen Meiksins Wood, "The Uses and Abuses of 'Civil Society'," *Socialist Register* (London: Merlin Press, 1990), pp. 60–84. For a more contemporary analysis see Manfred Bienfield, "Structural Adjustment and the Prospects for Democracy in Southern Africa," in David B. Moore and Gerald G. Schmitz (eds), *Debating Development Discourse: Institutional and Popular Perspectives* (New York: St Martin's, 1995).
20 For further critique of civil society in Africa, see Thomas M. Callaghy, "Civil Society, Democracy, and Economic Change in Africa" and Robert Fatton Jr, "Africa in the Age of Democratization: the Civic Limitations of Civil Society," *African Studies Review,* **38**, 2 (September 1995), 67–126.
21 Stephen N. Ndegwa, *The Two Faces of Civil Society: NGOs and Politics in Africa* (West Hartford, CT: Kumarian Press, 1996); Muigai, "Ethnicity."
22 Jennifer A. Widner, *The Rise of a Party State*; see also Ndegwa, *The Two Faces of Civil Society,* and Muigai, "Ethnicity."
23 Maria Nzomo, "Empowering Women for Democratic Change in Kenya: Which Way Forward?," in Ludgera Klemp (ed.), *Empowerment of Women in the Process of Democratization: Experiences on Kenya, Uganda and Tanzania* (Dar es Salaam: Friedrich Ebert Stiftung, 1994).
24 Ndegwa, *The Two Faces of Civil Society,* pp. 109–18.
25 *Daily Nation* (Nairobi), 16 July 1990, p. 1.
26 Joel Barkan, "Divergence and Convergence in Kenya and Tanzania: Pressure for Reform," in Joel D. Barkan (ed.), *Beyond Capitalism vs. Socialism in Kenya and Tanzania* (Boulder, CO: Lynne Reinner, 1994), pp. 1–45.
27 *Target Magazine* (Nairobi), 15 September 1992, pp. 1–2.
28 Barkan, "Divergence and Convergence," p. 38.
29 *The Weekly Review* (Nairobi), 23 October 1992. After a short absence from the cabinet Mr Biwott is now back as Minister in the President's Office.
30 *Daily Nation* (Nairobi), 15 August 1995.
31 International Republican Institute (IRI), *KENYA: The 29 December 1992 Elections* (Washington, DC: IRI, 1993).
32 *Daily Nation* (Nairobi), 3 September 1992, p. 1.
33 *Daily Nation* (Nairobi), 21 May 1992.
34 *Daily Nation* (Nairobi), various issues, May–July 1997.
35 Ali. A. Mazrui and Michael Tidy, *Nationalism and New States in Africa* (London: Heinemann, 1984).
36 See, for example, Colin Leys, *Underdevelopment in Kenya: The Political Economy of Neo-Colonialism* (Berkeley, CA: University of California Press, 1975). To explain Kenyan politics of the post-independence period, Leys pointedly moves away from ethnic explanations to that of structural underdevelopment on the basis of global capitalism.
37 Muigai, "Ethnicity," p. 161.
38 Diane Hunt, *The Impending Crisis in Kenya: The Case for Land Reform* (Aldershot: Gower, 1984).

39 Steven Langdon, *Multinationalism Corporations in the Political Economy of Kenya* (New York: St Martin's, 1981).
40 "Kenya: The Politics of Repression," *Race and Class*, Special Issue (London: Institute of Race Relations), **24**, 3 (1983).
41 For a detailed discussion on these matters please see Leys, *Underdevelopment in Kenya*; Hunt, *The Impending Crisis in Kenya*; Langdon, *Multinationalism Corporations in the Political Economy of Kenya*; *Race and Class*, Special Issue.
42 Gurushri Swamy, "Kenya: Patchy, Intermittent Commitment," in Ishrat Husain and Rashid Farugee (eds), *Adjustment in Africa: Lessons from Country Case Studies* (Washington, DC.: World Bank), p.13.
43 Ibid., p. 195.

5
The Incomplete Democratic Transition in Uganda

Susan Dicklitch

Introduction

The proverbial winds of change that swept over sub-Saharan Africa in 1989 can be considered as part of a "third wave of democratization" that has profoundly affected the African continent. Although Africa has been generally described as having "infertile" soil for the successful implantation of democracy, multi-party democratic change is occurring. For example, in 1989, 39 of the 45 sub-Saharan African countries had authoritarian forms of rule,[1] whereas, by early 1995, 31 of the 45 sub-Saharan countries held democratic presidential or parliamentary elections.[2] There have, however, been significant democratic reversals, and Africanists are now questioning the sustainability of these traditions and lamenting their lack of consolidation. In several cases economic and political liberalization and the "rebirth of civil society" have accompanied this dramatic change of events. But there is little conclusive evidence that economic and political liberalization, and the existence of a civil society, will necessarily lead to democratic consolidation.

Uganda offers important insights into the relationship between economic liberalization and democratization. Although the International Monetary Fund (IMF) and the World Bank consider Uganda to be one of the "economic miracles" of sub-Saharan Africa, Uganda is far from consolidating a democracy. This chapter critically examines several non-governmental organizations (NGOs) within civil society and their relationship to and role in the process of economic liberalization and democratization. This case study illustrates the co-optive tendencies of the National Resistance Movement (NRM) regime and the weakness of the non-state actors *vis-à-vis* the state, and thus helps to dispel the myth of a strong correlation between economic liberalization and the growth of civil society and democratization.

Economic liberalization, civil society and democratic transition

A growing body of revisionist literature on civil society argues that a causal relationship cannot be established between the growth of civil society and the transition to and consolidation of democracy.[3] Economic liberalization in sub-Saharan Africa *vis-à-vis* structural adjustment programs (SAPs), has led to a "privatization of both development and democracy."[4] Some would go as far as to argue that SAPs and economic liberalization have in essence created "international bantustans."[5]

> Governments and political processes have thus been emptied of political content. Governments have become executive agencies that implement predetermined, sound policies devised by technical experts. This has made it much easier to the IFIs [International Financial Institutions] (and the hegemonic powers) to support democracy in the developing world, since it is of no great threat to give people the democratic right to elect governments that have no effective power over social or economic policy. Indeed, under such conditions the attainment of democratic rights may turn out to be a blessing that makes it easier to divide and rule.[6]

It is thus important to note the political implications of economic liberalization.

The relevance of civil society to democratization

Civil society theoretically plays an important role in democratic transition and consolidation through empowerment, education, advocacy, and watchdog functions. It is becoming obvious, however, that not all civil societies are "constructive for democracy."[7] There are three main problems associated with civil society and democratization in Africa: the historical impediments to the vibrant growth of a strong and democratic civil society; the weakness and lack of democratic rule of many associations within civil society; and regime impediments. Although these bottlenecks are closely intertwined, this chapter concentrates mainly on the last two impediments.

Background to chaos

Uganda is perhaps most infamous for the bloodshed that characterized much of its post-colonial history. That bloodshed left significant colonial legacies. For example, the British policy of preferential treatment to the largest ethnic group, the Baganda, served to inflame anti-Baganda nationalism, regionalism and politicized ethnicity. The British practiced indirect rule through the Baganda ethnic group, whereby the Baganda chiefs became colonial middlemen and bureaucrats while other ethnic groups, particularly in the north, were used as labor and army reserves. Ugandans were thus conditioned by British colonial policies to think along ethnic lines. This was clearly reflected in post-colonial political life. The "ethnicization" of the political arena served to heighten suspicion and increase the zero-sum nature of political engagement. Because of this infiltration of ethnicity into politics, the securing of a semblance of authority and stability depended on patronage, coercion, repression and the support of the politicized military. In this context, institutional legitimacy was illusive, necessitating a further reliance on forced compliance. Instead of uniting Ugandans at independence, the nascent political parties served to magnify ethnic, regional and religious differences.

The importance of ethnicity in undermining the democratic potential in Uganda can be gauged not only by the degree to which it creates social disunity, but also by the degree to which it promotes the development of a politically weak and "unstable" state.[8] The lack of "unifying nationalism" undermined the creation of a viable nation-state. What emerged from this pot-pourri of ethnic rivalry and conflicting nationalism was a "fictitious state" which satisfied only the basic requirement for statehood – international recognition of territorial sovereignty – but lacked the most vital ingredient – internal legitimacy.[9] Loyalty, then, was expressed mainly to one's ethnic group and was rarely extended to the alien state entity. Future rulers of Uganda did not rely on the legitimation of political rule through institutionalization, but relied on force, repression and manipulation to secure often self-serving rule. Thus, coups, counter-coups, and civil war became the instruments by which an ethnic group would attempt to secure its dominance of the political system. This is illustrated by the fact that, since 1962 alone, Uganda has experienced five violent overturns of power.

Perhaps the most infamous of these overturns of power was the regime of Idi Amin Dada, CBE (self-named Conqueror of the British

Empire). During Amin's rule (1971–9), terror was used as "an instrument of domestic policy"[10] and accounted for the approximate 300 000–500 000 deaths.[11] The state under Amin lost all predictability and served mainly to advance the shifting interests of those in power: mainly Amin's ethnic group, the Kakwa and the Lugbara from the northwest (West Nile Region). The Acholis and the Langis[12] were the main targets of Amin's repression, although no one was really secure.

Nor was the economy secure. Part of Amin's initial popularity among Ugandans rested on his promise to Africanize the Ugandan economy. Hence, in 1972 he ordered the expulsion of all Asians (Ugandan citizens and non-citizens alike) from Uganda in 90 days. With the expulsion of over 49 000 Indo-Pakistani (Asian) merchants, manufacturers and civil servants, Amin virtually eliminated the whole Asian middle class.[13] These Asians (along with the foreign multinational subsidiaries which were also nationalized by Amin in 1972) had provided the foundation for industrial capital in Uganda as well as Uganda's domestic capitalist class based on the *dukawallah* (small Indian shopkeeper).[14] Amin's "economic war" allowed him to replace the Asians with his own kinsmen and military men. They squandered their newfound wealth and helped to destroy the once-productive Ugandan manufacturing and commercial base.

Amin's reign of terror set the stage for future regimes with a concomitant deepening of ethnic, religious and political schisms. Amin's regime was toppled in 1979 by the Ugandan National Liberation Army (UNLA) with the help of the Tanzanian army. Several transitional regimes followed, until Obote took back power in a questionable electoral victory in 1980. The Obote II regime was followed by a military coup headed by Ancholi army officers in 1985 which was overthrown by the guerilla National Resistance Army (NRA, the military wing of the National Resistance Movement, NRM) in January 1986. Yoweri Museveni, the head of the NRA/NRM, has retained power since the NRA/NRM takeover in 1986. He was recently elected President of Uganda in democratic elections held in 1996 and should carry Uganda into the twenty-first century.

The general violence and political repression that flourished in Uganda served to undermine the development of tolerance, national unity and trust – attitudes and values that are necessary for the development of a democratic civil society and polity. One of the more serious consequences of this situation was a clear "ethno-religious hemorrhaging" of civil society.[15] Virtually every family in Uganda lost a loved one in the ethnic violence that enshrouded Uganda for almost two decades. The willingness of average Ugandans to engage in political activity began to re-emerge with the development of the NRM system of Resistance Councils/Committees (renamed Local Councils in 1996). But the psychological impact of years of civil war affected many Ugandans who have become suspicious not only of politicians but also their fellow countrymen/women.

Beyond the chaos: the political arena and political liberalization

But the future for Uganda is no longer as grim as its past. This is mainly due to the economic successes of the NRM regime in power since 1986. The NRM regime has fully accepted market liberalization, fundamentally changing its economic policy orientation in 1987 for one bordering on Marxist revolutionary rhetoric to an almost complete embrace of economic liberalization. It has also stabilized the political arena, although this is being distracted by civil war in the north and northeast, with spillovers from the conflicts in surrounding countries.

Central to the NRM's political and economic strategy is its Ten Point Program.[16] Overall, the Ten Point Program seeks to establish popular democracy, security, national unity and the economic rehabilitation of Uganda. The NRM claims that this system is an alternative to multiparty democracy, based on a consensual, participatory, grassroots system of Local Councils (formerly labeled Resistance Councils). This is a hierarchical system based on five main levels of participation; from the Village Level (LC-1) to the District Level (LC-5), with the National Resistance Council and the National Executive Committee at the apex. This system gives the illusion of grassroots participation from the bottom up, but essentially ensures that political participation is filtered through the National Resistance Movement system, with directives often originating from the top down.

Under the banner of "no-party" politics the NRM further attempts to incorporate and control real and potential opposition from, for example, political parties, trade unions, NGOs, the media and some religious organizations. Although political parties are not officially banned, and they are allowed to have party headquarters, the regime prevents them from holding political rallies or campaigns. This initially stems from the "gentleman's agreement" that the NRM struck with the existing political parties in 1986: the Uganda People's Congress (UPC); the Democratic Party (DP); the Conservative Party (CP); and Museveni's former party, the Uganda Patriotic Movement

(UPM). The justification for such an agreement was based on the NRM regime's assertion that the devastating wars that proceeded NRM rule were due to sectarianism that stemmed from political parties manipulating the ethnic card in order to obtain power and destabilize the political arena. Without the nuisance of sectarianism the NRM could proceed to build national unity. Opposition political party members were invited to join the broad-based NRM which was not, according to the NRM, a political party but a "movement" embracing all. Thus political parties exist, but their powers are significantly circumscribed.

The promulgation of the new Constitution of Uganda, in November 1995, resulted in part with the implementation of Article 269 which effectively banned political party activities and multi-partyism until a referendum to be held in the year 2000. Thus political parties are still not formally banned (although their activities are now constitutionally limited), but multi-partyism is until the year 2000.

Popular participation is thus filtered through the structures of the LC system. The state considers autonomous associations that are political as "sectarian" and thus restricts them. This is the case with the Uganda Human Rights Activists (UHRA) and the detention of its Executive Director in 1987, and the temporary banning of the National Organization for Civic Education and Election Monitoring (NOCEM) in civic education during the 1995 Constituent Assembly elections. The regime, advocating a broad-based and inclusionary mode of participation, attempts to bring organizations that are "too political" in line with the NRM. As Oloka-Onyango observes, "the expression of views that do not correspond to the dominant party line are labeled as divisive and characterized as criminal rather than political."[17]

The above co-optive and sometimes repressive tendencies of the NRM regime are softened by a carefully formulated populist image with Yoweri Museveni, the charismatic "liberator of Uganda," at the helm. Museveni and his NRM regime offer Ugandans an alternative to the civil strife incited by the traditional political parties. Fighting corruption and eliminating the misuse of power, establishing popular democracy, restoring security and consolidating national unity and defending national independence are some of the related populist goals of the NRM.

Recently, the NRM regime has instituted several political liberalization initiatives. These include the holding of presidential and parliamentary elections in June and July 1996 for the first time since 1980, the drafting of a new constitution by a mainly elected constitutional commission in 1994 that was debated and passed by the constitutional assembly in November 1995, and the establishment of an Office of the Inspector-General of Government (IGG), the Ugandan Human Rights Commission and the Human Rights Desk in the Ministry of Justice.

In several cases, however, these instances of political liberalization are flawed. For example, although international observers declared the presidential and parliamentary elections as mostly free and fair, there were several pre-election inconsistencies and irregularities.[18] Many diplomats and opposition leaders maintained that the restriction of only one month of campaigning and the ban on political party activity gave Museveni such a heavy advantage that he could not lose the 1996 elections. In particular, the NRM used scare tactics to intimidated voters, like advertisements in the *New Vision* which suggested that one of the opposition presidential candidates, Paul Ssemogerere, was going to bring back Obote and the slaughter associated with his rule.[19] These advertisements were especially prevalent in the Luwero Triangle. In addition, Museveni supporters verbally and physically harassed, and even stoned and stampeded, the supporters of the two opposition candidates, Paul Ssemogerere and Joshua Mayanga, at numerous political rallies. Because of this, the Inter-Party Forces Committee (IPFC) boycotted parliamentary and local council polls in June 1996. After decades of civil strife it is perhaps no wonder that Ugandans chose the "safe" option of continued NRM rule.

Despite the historical weakness and divisiveness of political parties in Uganda, the political parties showed a considerable degree of unity during presidential and parliamentary elections, vetting one presidential candidate for the Inter-Party Political Forces Committee: Paul Ssemogerere. This was possible, not necessarily because the multipartyists (Democratic Party and Uganda People's Congress) were true democrats, but because they were unified in their desire to remove Museveni from the presidential office.

The establishment of the IGG, the Human Rights Commission and the Human Rights Desk are important moves in a democratic direction, but for the most part these organizations are plagued by underfunding and bureaucratic limitations. Although there are still recurring human rights abuses in Uganda, by government warring rebellious forces throughout the country, the situation in 1997 is much improved compared with that of the 1970s. However, Uganda was only given a human rights rating of 46 percent.[20]

Clearly, political liberalization in Uganda has not been matched with the fervor of economic liberalization. Museveni has said that he did not believe that Africa was ready for "full democracy" and likened

African societies to being at the "caterpillar stage" while other developed societies were like "butterflies." Implicit in Musevini's statement is the suggestion that there are tradeoffs to be made between economic growth and stability and a long-term commitment to democratization. Western donors and the IMF and World Bank have not pressured Uganda to adopt greater political liberalization, mainly because Uganda has so successfully and wholeheartedly implemented SAPs.

Uganda's economic recovery

When the NRA/NRM[21] took over Uganda, their economic policies were considered to be leftist-oriented and anti-West. In a quick turnabout in 1987, and with wide external support, the NRM regime adopted its first Economic Recovery Program (ERP). It targeted "strong private investment response and export diversification."[22] Between 1987 and 1994 the IMF disbursed about US$446.3 million to Uganda, while the World Bank gave approximately US$1.4 billion.[23]

Policies of structural adjustment have been implemented with varying degrees of success. For example, Uganda's shilling was devalued by 77 percent in May 1987 with subsequent minor devaluations and a devaluation in 1989 of 41.2 percent. In addition, there were price liberalizations on producer prices, including a 182 percent rise in the coffee price paid to farmers, and petrol prices were increased to reflect all landed and distribution costs. Approximately 20 percent of 200 000 civil service positions were retrenched, and there was a demobilization of 33 000 soldiers.[24]

Perhaps most importantly, from an economic liberalization perspective, "since 1987, trade and foreign exchange reforms have resulted in a substantial reduction in the anti-export bias."[25] The privatization program which began in 1993 under World Bank auspices has resulted in the selling-off of 30 enterprises to the tune of US$98.57 million.[26] Inflation has fallen from 250 percent in 1987 to 5 percent in 1996, while government employment has been slashed by half.[27] These policies follow a wider policy of decentralization, with the aim of regional districts assuming responsibility for as much as 80 percent of the delivery of services.[28]

Foreign investment has also increased with some 1400 new investment projects in 1991 for US$2.12 billion,[29] and the Indo-Pakistani Asians have returned to invest in Uganda with the introduction of Museveni's Departed Asians Property Custodian Board, although the number returning is far smaller than the 70 000 in 1971 who owned 90 percent of the country's business.[30]

Privatization has not occurred, however, without its challenges. For example, Parliament suspended the sale of parastatals in 1993 because of public pressure to the perceived "giving away" of the parastatals. The concern was that foreigners and elite figures within the current regime were buying the privatized parastatals at very low cost. Council members (CMs) in the National Resistance Council (NRC) criticized the NRM for the manner in which they were selling off public enterprises. They argued that the interests of Ugandans were not being protected during the privatization exercise.[31] Privatization was later restarted with parliamentary approval.[32] Despite this setback, Uganda has the sixth-fastest-growing economy in sub-Saharan Africa (1996), and is being praised by the International Monetary Fund (IMF) and Western donors as a "role-model" for Africa.[33] As an indication of the degree to which Uganda has been considered a "star performer in Africa," the IMF gave its stamp of approval for debt forgiveness of US$338 million. The World Bank also provided US$75 million in grants in 1998.[34]

This impressive economic "recovery" has been overshadowed by a growing total debt of US$3.5 billion to multilateral creditors.[35] The national debt ratio outstrips the national income by 110 percent. Its success follows over US$500 million in financial assistance to Uganda from the IMF over several years. The level of external debt has thus risen substantially in Uganda under the NRM regime. Although the Uganda economy is making a slow recovery, according to the IMF and the World Bank, this recovery seems far away for ordinary Ugandans who are bearing the brunt of the austerity programs.

Poverty is pervasive in the rural areas and the north, and the urban–rural gap is widening.[36] Famine has even been reported in several districts in Uganda in the southwest (Kasese), the north (Kitgum and Gulu, with over 700 000 facing severe food shortage)[37] and the east (Kgana, Kamuli, Tororo, Mbale, Pallisa and Kumi).[38] The government has projected that approximately 19 districts will be affected with over 2.1 million people likely to be hit with famine.[39] This famine can be attributed to insurgency and instability in the north (mainly the Lord's Resistance Army (LRA) and the West Nile Bank Front (WNBF), and rebels penetrating Ugandan borders, as well as drought. As a result, Uganda continues to have one of the lowest per-capita incomes in the world, approximately US$220.[40]

Economic liberalization and the impact of SAPs have mobilized a response from civil society. Local NGOs have joined the Structural Adjustment Participatory Review Network (SAPRIN), an international pressure group "agitating against the biting effects of the World Bank

SAPs."[41] There have been strikes over a new value-added tax (VAT), as well as railway workers' and medical workers' strikes. These strikes have been handled by the government in similar ways: police intervention to disperse the strikes, arrests of leading union officials and threats of revoking licenses.

The Ugandan Railways Corporation workers went on strike to protest their lack of payment of three-month salary arrears amounting to 4.5 million Ugandan shillings. The government reacted swiftly by sending in anti-riot police wielding batons and equipped with tar gas to "beat up" the striking workers.[42] Similarly, the government deployed police throughout the country with the 30 September 1996 VAT strike. This strike was the biggest non-political protest in Uganda's recent history.[43] Spearheaded by the Uganda Importers and Exporters Association (UGIETA), the strike was in protest against the 17 percent VAT introduced in June 1996. Strikers wanted to reduce the VAT to 10 percent. Museveni threatened to revoke the licenses of striking traders if they did not return to work.[44]

As noted above, the opening provided by economic liberalization has not necessarily represented increased avenues for participation within civil society. The regime has chosen to clamp down on dissent by issuing warnings to those groups who do not conform to NRM economic and political policies. There have been attempts by various groups to petition the state for change, but they have in most cases resulted in state co-optation or repression.

Civil society and the NRM regime

Political parties

Political parties almost everywhere have traditionally been key actors in the political arena, and crucial for pressuring governments to become legitimate, accountable and responsive to popular pressures. Although the activities of political parties are circumscribed, they have played an important role in bringing key political issues to the forefront. For example, they strongly petitioned against the proposed NRM use of the NRC as the debating organ for the Draft Constitution, they have criticized the five-year extension of the NRM regime in 1990, and they have been vocal critics of the IMF and World Bank structural adjustment programs in Uganda. They also vehemently opposed the proposed national referendum to decide movement versus multi-party politics in Uganda and criticized human rights abuses arising from UPDF sweeps of rebel areas in the north and the northeast.[45]

The NRM has reacted strongly against most political party activity. On one occasion Museveni even warned the political parties that "[i]f you want to bury, go ahead with the rally."[46] He apparently instructed the police to "use reasonable force to disperse any party rally held in Kampala or anywhere else in the country."[47] In another incident two UPC members were arrested by the police when they attempted to attend a UPC meeting in Mbale, headed by the General Secretary, Cecila Ogwal, in April 1993. The Deputy District Administrator (acting on the advice of the President), told Ogwal that the meeting was illegal.[48] Earlier in the NRM period, government troops raided the headquarters of the UPC in Kampala and charged the acting UPC leader, Paulo Muwanga, and several other UPC members, including Henry Bwanbule, Andrew Kayiira, David Lwanga, and Evaristo Nyzazi, with conspiracy to overthrow the government of Museveni. Several of these men were cabinet ministers and university lecturers.[49]

In addition to brute force, the NRM has also used detention to silence political party opposition. For example, in March 1994 the UPC National Chairman, Haji Badru Kendo Wegulo, and a number of other UPC officials were arrested by the NRM and charged with sedition for printing, publishing and offering for sale a book entitled *The 1994 Constituent Assembly Elections, the UPC Position*.[50] The DP has also had several officials charged with sedition and detained, including the Secretary-General of the DP, Robert Kitariko. In addition, 30 Uganda Young Democrats (UYD) were allegedly arrested by the police under the false pretext that they were engaging in "subversive activities." Thirty-four others were apparently on a police wanted list.[51]

State repression of political party rallies, political meetings[52] and so on, helps to undermine the long-term development of a consolidated democratic (i.e. other than ethnic or religious) foundation from which the political parties could launch viable alternatives to NRM rule. Political parties may be significantly weakened by internal bickering and in-fighting, but state restrictions further limit the effectiveness of political parties in Uganda.

The media

The media is also an important non-state actor. It has always been regarded as an important guarantor of people's rights. It exposes cases of corruption, of human rights abuses and other regime transgressions. It is the "watchdog" of the regime, *par excellence*. With over 40 000 newspapers in circulation daily, Ugandans, at least in the urban areas, are well informed about politics and corruption in Uganda.[53] Comparatively speaking, the Ugandan media has been one of the strongest in Africa. Under the Amin and Obote regimes its strength was eroded, but since the NRM came into power there has been a resurgence of media activity.

Although much more open and critical journalism has been allowed under the NRM regime, the regime seems to be tightening its control over the media and what is considerable permissible journalism. Each political party has its own newspaper (i.e. the DP has *The Citizen*, the UPC has *The People*, and the government has its own paper as well, *The New Vision)*. There are several other significant newspapers, such as *The Weekly Topic* (which collapsed in 1994), and *The Crusader* which started with ten senior journalists who left *The Monitor* in 1995.[54] The biggest newspaper by far is *The New Vision*, which has an editorial staff of 60 with a circulation of 25 000. However, the NRM has increasingly clamped down on press freedom on several occasions. In fact, Freedom House has rated Uganda as having one of the "worst" scores on press freedom.[55]

Almost from the onset of NRM power in 1987, Museveni warned journalists that they would be put under "preventative detention if they write stories which, in the view of the government, discredit the NRA."[56] Similarly, in 1989, Museveni stated that "two types of criminals interest me: journalists and common criminals. They think they are above the law."[57] Numerous arrests of journalists on several trumped-up charges, including defamation, treason and sedition, have to a certain extent characterized NRM–media relations. For example, the editor of the daily newspaper, *Etop*, Charles Ocan, was charged with criminal defamation (the charge was eventually dropped), while Saul Katabarwa of *The Guide* and John Kyeyune of *The Citizen* were also arrested. The editor-in-chief of *The Monitor*, Wafula Ogutu, was charged in October 1994 with four counts of "publishing false news," although he was freed after those charges were dropped by the Director of Public Prosecution.[58] More seriously, Akuta Too, a journalist, was arrested and subsequently died upon his release from prison.[59] Similarly, Hussein Mus Njuki, the editor the Islamic newsletter, *Assalaam*, was arrested by police for printing articles criticizing Museveni, and died after being in police custody. The editor of the bi-weekly *Shariat* newsletter, Hanuna Kanaabi, was also charged with Njuki for allegedly "belittling" Museveni and wishing him death in an article which said that Uganda would never have peace until Museveni died.[60] Kanaabi was fined US$1200 and given a five-month jail sentence. The Uganda Journalists'

Association claimed that the sentence was politically motivated to muzzle critical journalists before the spring 1995 elections were held.[61]

Another journalist, Vivian Asedri, was detained after traveling to Lira following the UPDF ban on journalists traveling to the area during army operations against rebels.[62] Also more recently, Lawrence Kiwanuka, editor of *The Citizen*, was arrested and charged with sedition in April 1995 for "publishing faulty information likely to cause disaffection with the state." Kiwanuka wrote an article accusing the Museveni government of "backing rebel groups in neighboring Kenya and Sudan."[63]

The case of Teddy Sezzi-Cheeye, editor of the *Uganda Confidential* newsletter, is also illustrative of NRM intolerance to criticism. Cheeye has exposed numerous cases of government corruption, sometimes involving high-ranking government officials. He has been charged with sedition several times for publishing libelous material against the Chief Justice of Uganda, and accusing Museveni and other high-ranking officials of corruption and the defamation of a prominent banker.[64] He was also charged, under questionable circumstances, with kidnaping a woman with the intent to harm her. The regime seems intent on muzzling the very vocal Cheeye. The detentions, arrests, harassments and deaths in police custody have led to a great deal of self-censorship on the part of the press.[65]

There have also been more sophisticated and covert attempts to limit the power of the press and create disunity within it. For example, the President tried to establish a rival body to the Ugandan Journalists' Association (UJA) in 1993. This rival group (composed mainly of the editors of the main newspapers) met with the President on a regular basis, where important state matters would be discussed, with the caveat that the discussions were not to be made public. This undermined the unity of the journalists (some had the President's ear while others were excluded) and undermined their diligence against the state.[66] This government interference further exacerbates the disunity within the media itself, and the internal disputes within the UJA. For example, most Ugandan journalists believed that the UJA president in 1991, James Namakajo, was linked to the NRM government.[67] There tends to be little solidarity among journalists, which overall limits the watchdog potential of the media.

To further undermine the strength of the media, in July 1993 the government implemented a decision to deny private newspapers public-sector advertisements. This was a move to strangle them economically for allegedly publishing government secrets and critical articles. All government departments and parastatals were ordered to cancel all advertisements that they had sent to the private press prior to the cabinet decision.[68]

A more recent attempt by the NRM government to silence the critical presses in Uganda is the Mass Media and Journalists Statute of 1995. The bill was originally tabled in the NRC on 16 April 1994. The statute,

> provides for the enrollment, registration, and licensing of practicing journalists in the country. It prohibits practicing journalism without a certificate issued by the Media Council. The council has been given the powers to refuse issuing practicing certificates to journalists of unsound mind, those who fail to comply with an order made under the statute and those who have proceeded against professional conduct.[69]

The Media Council (named the National Institute of Journalists of Uganda, NIJU) is composed of 11 members, appointed by the Minister of Information, in addition to a Disciplinary Committee composed of five members all appointed by the Minister as well. The statute "suspends publication of any government classified matter without the approval of the Minister of Information."[70] The statute is quite comprehensive, in that it affects all newspapers, radios, televisions, cinemas and video operators. One Ugandan journalist described the original bill as a " 'hair-raising proposition' that essentially indicates that the government has declared war on journalists with all variables staked in its favour."[71] NIJU started registering journalists in March 1996.

The 1994 bill and subsequent statute follow a 1991 bill that the regime attempted to push through the NRC. It would have required journalists to register with a central government body, and would have given ministers the power to stop certain articles from being published. This bill was vehemently fought by journalists and was finally dropped from the NRC.[72]

Print media has a limited impact on disseminating information, due to its mainly urban distribution (whereas the majority of the population in Uganda live in rural areas), and to the fact that it is more expensive than other forms of information and literacy issues. Radio has thus risen in importance in Uganda. As a consequence, the Minister of Information in the NRM regime introduced an Electronic Media Bill in March 1996. It regulates radio broadcasting stations and other electronic media, and provides for the licencing of TV and TV sets.[73]

This latest bill is just another step in increased NRM regulation and control of the media in Uganda. Government action to control and manipulate the Ugandan media gives a clear message to political critics of the NRM regime: criticism will not be tolerated. For example, Museveni instructed the security agencies to "monitor all radio broadcasts and to identify those who incite the public against government policies."[74] Nevertheless, the Ugandan media continues to be one of the strongest advocates for regime accountability and openness in Uganda. It has continued to print articles on corrupt government officials, on the continued instability in the north and the northeast, and on human rights abuses by the regime.

Non-governmental organizations: the case study of NOCEM

The growth of non-governmental organizations (NGOs) in Uganda since 1986 has been massive. Self-help groups/associations have always existed in Uganda, and became especially important for survival during the virtual collapse of the state in 1970s until the mid-1980s. Since NGOs exist in such large numbers in Uganda, the government cannot simply ignore them. How it chooses to deal with them, however, is not so clear-cut. The regime realizes that it needs NGOs, particularly in terms of poverty-alleviation and service provision, but it could do without the more troublesome, vocal NGOs that have either a direct or indirect political agenda. The government has applied both the carrot and the stick approach in dealing with the potential and actual opposition. The NRM regime has relied most heavily on the co-optation of individuals and groups under its broad-based rule, but state repression often has been used with varying degrees of intensity to ensure that things do not get too far out of line with the goals of the regime.

There seems to be a tacit understanding as to the limits of non-compliance with NRM policy that the NRM will tolerate. This may explain why some NGOs prefer to remain apolitical and non-confrontational. For example, the former executive director of the Uganda Human Rights Activists (UHRA), Lance Seera-Muwanga, was detained without trial for 13 months in 1987 under the Public Order and Security Act because he criticized the NRM's human rights record in the north. This significantly muzzled the future aggressiveness of the organization, which now channels its human rights concerns through the office of the Inspector-General of Government. UHRA now views itself as having a more "educative role in human rights, rather than being a pressure group."[75]

A more moderate warning was given to the UHRA by the District Administrator (DA) of Entebbe, when he addressed a paralegal workshop for UHRA at Entebbe. The DA called on the UHRA to serve as mediators and peace creators and avoid political gains. He commended the work of UHRA but warned them not to "misuse their forum for political gains."[76] The DA's warning was indicative of the NRM's willingness to support the UHRA's activities as long as they did not get too political or, more pointedly, critical of the NRM regime.

Similarly, the Foundation for African Development (FAD), an organization that has been closely associated with the Democratic Party (DP), has been accused by the President of not heeding the NRM's advice to let the interim period be a time of "national recuperation" allowing the population to chart their own future through the constitution-making process.[77] FAD was accused by the President in a closed NRC session of "involvement in activities to destabilize Uganda's politics."[78] In another incident, the then Prime Minister Kisekka stopped a FAD seminar on proposals for Uganda's new constitution because it was deemed premature and did not fit the NRM timetable. These NRM warnings and actions serve to remind NGOs of their subservient and tenuous position in Uganda. Anything that is deemed critical is classified as sectarian and destabilizing.

Consequently, many NGOs shun political activities altogether. The International Association of Women Lawyers, Uganda Branch (FIDA), for example, does not focus on politically sensitive issues such as the detention of women in military barracks, but prefers to remain in areas that will not bring it into confrontation with the regime.[79] Similarly, Action for Development (ACFODE), a prominent and important women's empowerment organization, vehemently denies and rejects any political orientation in its activities. [80]

Within this scenario, NGOs fit into the LC system as facilitators of NRM objectives, not as alternative sources of accountability or power. NGOs will be tolerated by the regime as long as they continue to function in an apolitical and non-confrontational manner. The degree of autonomy and power that NGOs have outside of the LC system is therefore limited, and the actual degree of latitude they has is conceded to them by the regime. The best illustration of this is the NRM reaction to the umbrella organization of NGOs: the National Organization for Civic Education and Election Monitoring (NOCEM).

NOCEM was made up of 14 indigenous NGOs, including the human rights organizations UHRA, Foundation for Human Rights Initiatives (FHRI), the Law Society of Uganda, FIDA–Uganda, the Uganda Journalists Association (UJA), the Islamic Information Centre and the Makerere Law Society, the Ugandan Christian Prisoners Aid Foundation (UCPAF), Sustain Africa (an environmental group), the Uganda Newspapers and Publishers' Association (UNEPA), the Uganda Federation of Business and Professional Women (UFBPW), the Youth Alliance for Development Co-operation (YADCO), Uganda Community Based Child Care (UCOBAC) and ACFODE (which later distanced itself from the organization). NOCEM was officially formed in August 1993 at a meeting called by the Uganda Law Society headed by the President of the Law Society, Mrs Solome Bossa. Its focus was non-partisan and non-profit while its mandate centered on civic education, and election monitoring of the Constituent Assembly elections held in May 1994 and the parliamentary and presidential elections in 1995. It was accredited by the Constituent Assembly Commission to carry out civic education and elections monitoring, and is funded mainly by the Danish Development Agency (DANIDA), the United States Agency for International Development (USAID), and the African-American Institute (AAI). NOCEM trained some 4054 civic educators (the volunteers mainly teachers and lawyers) and monitors in 20 pilot districts in Uganda, and had some 5000 poll watchers covering polling stations nationwide.[81]

Rumors quickly spread that NOCEM was "partisan" in its civic education; that is to say, it was not pro-NRM. Pressure to ban NOCEM came from top levels in government, including Eriya Kategaya, the First Deputy Prime Minister and the National Political Commissar. He found it "completely unacceptable" that a number of NOCEM co-ordinators' political identities were UPC. In fact, on the eighth anniversary celebration of the NRM victory, President Museveni publicly singled out the NOCEM administrator (Macarius Asubo), as well as the NOCEM organization, as "criminal." In essence, NOCEM was banned from civic education, although it was not banned from participating in election monitoring.

The case of NOCEM illuminates the limitations that NGOs are faced with daily. Even though NOCEM may have contained a number of UPC supporters, its work as civic educators and election monitors was crucial for building a foundation for openness, dialog and compromise. The regime's reaction to NOCEM is therefore troubling with regard to the overall prospects for genuine political participation and openness in Uganda.

Uganda's troubled road to democracy

Although Uganda is considered an African success story with regard to its aggressive implementation of economic liberalization, political liberalization and democratization have preceded at a much slower pace. Civil society has awakened to respond/react to some changes from above (implemented by the regime), but there is still a long way to go before civil society can become proactive to state policies and initiative. Organizations are allowed to exist in Uganda as long as they solely perform a "gap-filling" role. NOCEM, as examined above, exists as an umbrella organization for NGOs within Uganda, but remains largely restricted in making an impact on regime policies. This is reflective not only of the internal limitations of the organization itself, but also of regime controls on the organization and its membership.

The NRM regime coopts and, when necessary, represses organizations and individuals who do not follow the NRM's schedule for economic and political liberalization. The no-party system, which is increasingly resembling a one-party system, controls the degree and pace of economic liberalization in Uganda, almost regardless of popular reaction. Although most Ugandans welcome the economic "successes" of the NRM regime, it seems that they also fear being too vocal about the negative aspects of these reforms. Uganda may be on the road to economic growth and restructuring, but there is a long trek ahead before the NRM embraces democracy in Uganda.

Notes

1 Claude Ake, *Democracy and Development in Africa* (Washington, DC: The Brookings Institution, 1996), p. 135.
2 United Nations, "Democratization," *African Recovery*, **10**, 2 (October 1996), 24.
3 Laura Mcdonald, *Supporting Civil Society: The Political Role of Non-Governmental Organizations in Central America* (New York: Macmillan and St Martin's Press, 1997).
4 Laura Macdonald, "Turning to Non-Governmental Organizations: Competing Conceptions of Civil Society in Latin America." Paper presented to the annual meeting of the Latin American Studies Association, Los Angeles, September 1992, p. 6.
5 David Himbara and Dagwood H. Sultan, "Reconstructing the Ugandan State and Economy: the Challenge of the International Bantustan," *Review of African Political Economy*, **63** (1995), 85–93.
6 Manfred Bienefeld, "Structural Adjustment and the Prospects for Democracy in Southern Africa," in David B. Moore and Gerald J. Schmitz (eds), *Debating Development Discourse* (New York: St Martin's Press, 1995), p. 108.
7 E. Gyimah-Boadi, "Civil Society in Africa," *Journal of Democracy*, **7**, 2 (April 1996), 108; Rene Lemarchand, "Uncivil States and Civil Societies: How Illusion Becomes Reality," *Journal of Modern African Studies*, **30**, 2 (1992), 177–91; Macdonald, *Supporting Civil Society*.
8 John Saul, "The Unsteady State: Uganda, Obote and General Amin," *Review of African Political Economy*, **5** (January-April 1976), 12–38.
9 Lemarchand, "Quelles Independances?," *Le Mois en Afrique* (February–March 1983), 19–37.

10 Dan M. Mudoola, "Communal Conflicts in the Military and its Political Consequences," in Kumar Rupesighe (ed.), *Conflict Resolution in Uganda* (Oslo, London and Athens, OH: International Peace Research Institute in association with James Carrey and Ohio University Press, 1989), p. 131.
11 Jan Jelmert Jorgensen, *Uganda: A Modern History* (London: Croom Helm, 1981), p. 314.
12 These ethnic groups represented Milton Obote's ethnic group.
13 Jorgensen, op. cit., p. 247.
14 Himbara and Sultan, op. cit.
15 Jorgensen, op. cit., p. 294.
16 The NRM "Ten Point Program" consists of the goals that the NRM developed during the People's Protracted War of 1981–5, that included the establishment of popular democracy as well as a list of nationalist unifying elements (NRM Secretariat, 1990, pp. 8–35).
17 Joe Oloka-Onyango, "The National Resistance Movement, 'Grassroots Democracy' and Dictatorship in Uganda," in Robin Cohen and Harry Goulbourne (eds), *Democracy and Socialism in Africa* (Boulder, CO: Westview Press, 1991), p. 131.
18 *New York Times News Service*, 10 May 1996.
19 The Luwero Triangle was infamous for the slaughter that occurred there. Milton Obote's forces massacred thousands of Baganda who were accused of supporting Museveni's NRA.
20 Charles Humana, "Uganda," *World Human Rights Guide*, 3rd edn (New York: Oxford University Press, 1992), p. 338.
21 The NRM is the political wing of the NRA (the National Resistance Army).
22 World Bank, *Uganda: The Challenge of Growth and Poverty Reduction – Country Study* (Washington, DC: World Bank, 1996), p. xi.
23 Robert L. Sharer *et al.*, *Uganda: Adjustment with Growth, 1987–94: IMF Occasional Paper 121* (Washington, DC: IMF, 1995), p. 2.
24 Ibid.
25 Ibid., pp. 3–5, xi.
26 Reuters, 11 February 1996.
27 *Newsweek* (Special Supplement), 20 January 1996.
28 World Bank, op. cit., p. xii.
29 "Uganda: Brits Top the List," *The Indian Ocean Newsletter*, 3 February 1996.
30 *Financial Times*, 5 February 1996
31 *New Vision*, 22 July 1992
32 Reuters, 11 February 1996.
33 Reuters, 21 February 1996.
34 Reuters, 23 April 1997.
35 United Nations, "UN–NADAF Mid-term Review: Country Profiles," *Africa Recovery*, **10**, 2 (October 1996).
36 World Bank, op. cit., p. 79.
37 *New Vision*, 23 November 1996.
38 *Uganda News*, 30 November 1996.
39 *New Vision*, 5 March 1997.
40 World Bank, op. cit., p. xi.
41 *The Monitor*, 3 March 1997
42 *New Vision*, 3 January 1996.

43 *East African*, 7–13 October 1996.
44 *Features Africa Network*, 4 October 1996.
45 *New Vision*, 27 July 1992.
46 *The Monitor*, 7–11 May 1993.
47 Ibid.
48 *New Vision*, 24 April 1993.
49 *Index on Censorship*, 1987, p. 44.
50 *New Vision*, 5 March 1994.
51 *The Monitor*, 16–18 October 1996.
52 Norm Olsen, Programme Director, USAID Kampala. Interview by author, 27 April 1993, Kampala, Uganda.
53 Apparently six to ten people on average read the same paper. Wafula Oguto, editor of *The Monitor*, interview by author, 21 January 1993, Kampala, Uganda.
54 Inter Press Service, 6 December 1995.
55 *The Monitor*, 10–13 May 1994
56 *Index on Censorship*, 1987, p. 39.
57 International Centre on Censorship, 1991, p. 55.
58 *New Vision*, 3 January 1995.
59 *The Monitor*, 13–17 August 1993, p. 9.
60 Agence France Presse, 31 August 1995.
61 Reuters, 28 December 1995.
62 *Weekly Topic*, 17 May 1991.
63 Reuters, 20 April 1995.
64 FBIS, 4 October 1993; *New Vision*, 20 February 1996.
65 *The Monitor*, 13–17 August 1993, p. 9.
66 James Okanya, Uganda Journalists' Association, interview by author, 13 April 1993, Kampala, Uganda.
67 International Centre on Censorship, 1991, p. 59.
68 Cabinet Minute No. 181 [CT 1993] of 14 July 1993; *The Monitor*, 27–31 August 1993, p. 1. This ban was lifted in 1997, *The Monitor*, 26 April 1997.
69 *New Vision*, 14 April 1994.
70 Ibid.
71 Ibid., 13 May 1994.
72 Confidential interview, 10 May 1993, Kampala, Uganda.
73 *The New Vision*, 6 March 1996.
74 *The Monitor*, 14–18 October 1996.
75 S. Bukenya Kasaato, Executive Secretary, Uganda Human Rights Activists, interview by author, 18 January 1993, Kampala, Uganda
76 *New Vision*, 5 May 1992.
77 Ibid., 28 August 1992.
78 Ibid.
79 Remy Kasule, Former President of the Uganda Law Society, interview by author, 10 May 1993, Kampala, Uganda.
80 Margaret Kikampikaho, Acting Executive Secretary, ACFODE, interview by author, 15 November 1992, Kampala, Uganda.
81 NOCEM, 1994, p. 3.

6
Structural Adjustment, Military Dictatorship, Civil Society, and the Struggle for Democracy in Nigeria

Funso Afolayan

Introduction

On 27 February 1999, Nigerians went to the polls to choose a new president. Former military Head of State (1976–79) and retired army General Olusegun Obasanjo, of the People's Democratic Party (PDP), defeated Chief Olu Falae, the joint candidate of the Alliance for Democracy (AD) and the All People's Party (APP) to emerge victorious. Obasanjo's election was the culmination of a series of elections put in place by General Abubakar, who succeeded General Sanni Abacha, following his sudden death on 8 June 1998, apparently from a nocturnal heart attack. The news of Abacha's death was received internationally with relief. In Nigeria it was welcomed with bizarre celebrations, as thousands of Nigerians poured into the streets of its major cities to jubilate over this most welcome demise of their ruler. In the Muslim city of Ilorin in central Nigeria the crowd of celebrants was estimated at about 10 000. Anyone familiar with the turn of events in Nigeria since 1993 was not surprised at this response.

On 12 June 1993 Nigerians went to the polls to elect a new president after nearly two and a half decades, aside from a brief intermission, of military rule. The election was generally peaceful. Domestic and international observers adjudged it as the most free and the fairest in the nation's history. When it became certain that the presidential candidate of the Social Democratic Party (SDP), Moshood Abiola, was poised to win by a large majority, the military government of General Ibrahim Babangida suddenly intervened. It nullified the results of the election and suspended the whole transition program. All over the country the nullification was condemned as an act of injustice, an assault to the moral integrity of the nation, a blatant denial and frustration of the collective will of the Nigerian people, and the worst manifestation of the arrogance and the irresponsibility of the Nigerian military dictatorship. Nigeria was thrown into a profound crisis from which it has not recovered. So vehement and relentless was the opposition that Babangida was soon forced to relinquish power to an interim government, headed by Earnest Shonekan. The new government was a mere front for the military. It lacked legitimacy and could neither restore order nor resolve the election nullification crisis. Its failure became the pretext for General Sanni Abacha, who was Babangida's Defense Chief, to seize power and abrogate the entire transition program.

Using Nigeria as a case study, and focusing on the 1993 presidential election, this chapter will examine the connections between structural adjustment, democratization and civil society in post-colonial Africa. The chapter argues that the pervasive state of economic crisis and debilitating poverty, the entrenchment of military dictatorship and the failure of democratization in Nigeria are closely related to, and in most cases the direct results of, the strangulating consequences of the World Bank/IMF-sponsored imposition and implementation of a structural adjustment program in the country. The fall-out of all these in the reawakening of civil society,[1] the resurgence of popular culture and the intensification of the struggle for civil liberty, protection of human rights and democracy will be examined.

The antecedents

To understand the perennial crisis of nation building in Nigeria one must first understand the nature of Nigeria as a political entity and a society. As a political entity Nigeria was a creation of British imperialism. It is made up of over 250 ethno-linguistic groups, who before the colonial era did not form a single political union. With varying degrees of sociocultural differences they were organized into numerous kingdoms, states and acephalous societies. Though there were considerable economic and cultural interactions, and the occasional overlapping of political frontiers, the different groups jealously guided and maintained their political and cultural autonomy. Forced together into a single union by the infrastructures, administrative convenience and political exigencies of colonialism, the various Nigerian groups have had to learn to live together as corporate members of a single nation, since the attainment of independence on 1 October 1960. For Nigerians it has been a most difficult and tortuous journey, filled with fits and starts, trials and errors, blood and tears.

In its geopolitical formation Nigeria can easily be divided into five major configurations. These are the predominantly Muslim Hausa-Fulani in the north, the partly Christian and partly Muslim Yoruba in the west, the overwhelmingly Christian Igbo in the east, the partly Muslim and partly Christian numerous ethnic minority groups in the northern middle belt and the mainly Christian minorities in the southern coastal areas.[2]

The competition for power has been mainly between the three dominant groups, the Hausa-Fulani, Yoruba and Igbo, who between them constitute about 60 percent of the 115 million people domiciled in the country. Political power has, however, remained solidly in the hands of the Hausa-Fulani, whose ruling elite (both military and civilian) has strenuously and successfully resisted every challenge to its monopoly of political power in the country. The northern ruling elite has been able to do this largely by adroitly courting the support of the minority groups in the north and south, while strategically keeping its most threatening rivals, the Yoruba and Igbo, divided. In 1993 the chickens appeared to have come home to roost. The strategy of divide-and-conquer failed to work. The party, National Republican Convention (NRC), dominated by the northern ruling elite was overwhelmingly defeated in the north as well as in the south. Its presidential candidate, Bashir Tofa, could not carry the vote even in his own state of Kano, the most populous Islamic city in the country. The presidential vote was a clear rejection by the majority of Nigerians, and most especially northerners, of the divisive and ruinous leadership of the northern-dominated ruling elite. For the northern ruling elite and their cronies to the south this defeat was a disaster. Abiola's victory signaled the beginning of a power shift from the northern ruling oligarchy. To accept it was to commit political suicide. After recovering from the shock of their defeat the NRC immediately reorganized and colluded with the military to validate the annulment of the election and pave the way for the perpetuation of military rule.

At the time of independence, and under the tutelage of the departing British officials, the emergent elite adopted the Westminster model of parliamentary democracy and a federalist structure of government. The new nation would, however, enjoy neither democracy nor federalism. The intra-elite struggle for power and control of economic resources, and their systematic and invidious manipulation of ethno-sectarian differences, soon brought the nation to the brink of disaster. In January 1966 the army struck to put an end to the indiscipline and madness of irresponsible civilian rule.[3] The military, however, could not arrest the tottering push toward disintegration. A 30-month civil war succeeded in keeping the nation together, while at the same time politicizing the army and entrenching it into the political fabric of the nation. Popular discontent with military lawlessness and unbridled corruption led to yet another shot at civil democratic government in 1979. The new civilian regime, led by Alhaji Shehu Shagari and made up mostly of the survivors of the old regime, soon excelled all its predecessors in the magnitude and brazen nature of its corruption. On the eve of the new year on 31 December 1983 the army, ever waiting in the wings, rolled out its tanks again and seized power.[4] The harshness and the authoritarian policies of the new regime headed by General Mohammed Buhari led to its termination. In July 1985, after spearheading a palace coup, the chief of Army Staff, General Ibrahim Babangida, assumed the leadership of the country with the significant title of President.

Babangida and the structural adjustment program

Whatever its eventual débacle, the Babangida administration took its self-imposed task of ruling the nation seriously. To its credit it initiated and embarked on many programs and projects to achieve its aims. The economy was the first area to receive immediate attention. Years of unbridled corruption, mismanagement, and mindless prolificacy had badly damaged the economy and impoverished the nation. The belated and haphazard austerity measures hastily introduced in 1982 were too weak to arrest the continuing economic decay and deterioration. The Buhari regime had firmly resisted the pressures of the IMF to implement its structural adjustment program conditions on trade liberalization, blanket devaluation and desubsidization without questions. Its overthrow was a welcome relief to the IMF. The new regime of Babangida knew that for its survival it must resolve the conflict with the IMF over the nation's external debt, then estimated at 19 000 billion. It set up an independent commission, charging it with the responsibility of conducting a genuinely free and open national debate on the desirability of receiving the IMF loan package. The debate was conducted over the airwaves, on the pages of the print media, in institutional and public seminars, in churches and mosques, in worker's unions and market-women associations' meetings, and in village and town assemblies.

If the government had hoped that the national debate would give it the necessary mandate to take the loan, it was grossly mistaken.

Everywhere the opponents of the loan carried the day. They argued convincingly that most countries often come out worse off than before they took the loan.

Consequently, on 12 December the government announced its decision not to take the IMF support loan in spite of pressures from its international creditors and the World Bank to do so. Nevertheless, the government went ahead and began to implement systematically the various conditionalities attached to the IMF loan without taking the loan. In spite of the pretension to the contrary, the new "home-grown" package was negotiated with World Bank officials and received their blessings.

Unable to resist the prodding of the IMF for a formal agreement, the government, without any public consultation, went ahead and took the IMF loan. Clearly the sharp decline in the price of oil in the world market, from US$30 a barrel to US$12 a barrel in 1986 influenced Babangida's ultimate decision.

The social and economic costs of structural adjustments

In July 1986 the Structural Adjustment Program (SAP) was introduced. It was followed in September by the Second-Tier Foreign Exchange Market (SFEM). This was to have the most far-reaching consequences as it began the gradual devaluation of the Nigerian currency, the naira ₦; a devaluation the IMF had been requesting since 1983 in order to allow the naira to find its level in a free market. This, it was believed, would discourage import while encouraging export. Unfortunately this did not happen. The value of the naira plummeted far beyond government control and far below the initial IMF expectations. From an exchange rate of US$1 to ₦2 in 1986, a dollar was being sold in the parallel market for ₦50 in late 1993. The value of the naira continued to plummet. By early 1998 one dollar was being exchanged for ₦85. Since 1989 the inflation rate had remained at a very hyper-level with prices tripling every year. By 1993 the country's external debt stood at about US$35 billion. The per-capita income fell from US$1500.00 in 1980 to US$395.00 in 1993 and has continued falling. Saddled with debt, declining productivity, and a crippling poverty, Nigeria, which in 1980 was one of the richest nations in the world, was classified by the World Bank as one of the poorest in 1995.[5]

Thus having adopted the harsh austerity measures imposed by its predecessors, the Babangida regime decided to push them to their IMF/World Bank-inspired logical conclusion by reducing state involvement in the economy and giving free rein to market forces. The massive devaluation and the high cost of borrowing led to an increase in production cost for industries, especially since most of them depended on external markets for their raw materials and spare parts. Prices of basic commodities became exorbitantly high and well beyond the real purchasing power of most Nigerians, who had to cope with the pangs of pay cuts and hyperinflation. Unemployment increased as many industries closed down, while the few surviving ones carried out massive retrenchment to save costs and stay afloat.

Life became hard and harsh for the majority of the population. As the economy contracted, the ensuing discontentment resulted in brain drain to the private sector and to other countries. Others who fled the country declared themselves as refugees in order to qualify for asylum in other countries.[6] Several sectors were devastated, including the health sector. By 1990 the per-capita allocation to health was US$0.62. As the most qualified doctors and other health workers joined the brain drain to Middle Eastern countries, most especially Saudi Arabia, Yemen, Oman and Kuwait, the hospitals became mortuaries. The sick avoided hospitals like a plague. The few who found the money to patronize the hospitals, and come back alive, often brought home new diseases contracted while on admission. Bread and milk disappeared from the tables of most Nigerians, except the rich few. The rising cost of living and unemployment meant poor diet, malnutrition, a lack of access to health care and a low resistance to diseases.

Available statistics present a gloomy picture of the impact of SAP on health care in the country. Ailments previously declared to have been eradicated in the country staged their comeback. Record cases of death from malaria increased from 1.2 million in 1984 to 1.8 million in 1988. The deteriorating sanitation and poor water quality increased the mortality rate from cholera, a water-borne disease, from 136 between 1979 and 1983 to 325 deaths between 1984 and 1988. The number of deaths from measles, a treatable disease, was 7480 between 1984 and 1988. In 1984 there were only five reported cases of death from yellow fever. This figure shot up to 599 in 1987 and 1531 in 1988. With no funding, the government's widely publicized propaganda slogan of "Health for all by the year 2000," soon became "Hell for all by the year 2000."[7]

Nevertheless, the era was not without a few positive developments. A new population census was successfully carried out. In March 1992 the President announced a new population figure of 88.5 million for the country. In August 1991 nine new states as well as 47 new local government areas were created to satisfy local demands for autonomy, and bring government and development nearer to the people. In December 1991 the seat of the federal government was moved from Lagos to Abuja, marking the formal inauguration of Abuja as the new federal capital. To ease some of the perceivable pains of adjustments the government took various measures. These included the establishment of Better Life for Rural Women (under the directorship of Maryam, the president's wife), the Directorate for Food, Roads and Rural Infrastructure (DFRRI), and the National Directorate of Employment (NDE).

However, most of these programs made limited impact. The gains of adjustment were eroded at the official level by the persistence of massive corruption, favoritism, enthronement of mediocrity, gross inefficiency and the failure to creatively adapt the IMF prescriptions to the sociopolitical realities of the country. The over a billion dollars invested in the programs only ended up enriching contractors and bureaucrats, who did not have to account for their failure to provide food, shelter, roads, and medicine to the rural population.

Within a few years the pangs of SAP produced a string of workers' strikes, student protests and market women's demonstrations. A series of violent riots broke out in the major urban centers in 1986, 1989, 1990 and 1992. Many of the riots were started by students. Other disgruntled elements, never in short supply in Babangida's SAP-sapped Nigeria, soon joined to swell their ranks. Most of the riots started as spontaneous explosions over specific acts of the government, such as sudden removal of petroleum and college tuition subsidies or, in one case, rumor of gross embezzlement of public funds by top military notchers.[8] While they caused much damage, especially to government properties, they were generally uncoordinated and, as such, vulnerable to the government's repressive measures.

Even the army was riddled with division. The lower ranks became disgruntled with the government's incessant calls for sacrifices while senior military officers and top governments' officials continued to live in affluence. Two attempted military coups in 1986 and 1990 have been linked with the general disenchantment with the economic policies of the government.[9] In most cases the riots were brutally suppressed, while scores of army officers were executed or incarcerated for the coups. Feeling insecure, with its initial popularity waning over the ravages of SAP, and under pressures from protesters at home and creditors from abroad, the government stepped up its suppression of labor unions, student organizations and the press. Laws were passed to stop further demonstrations against SAP or against the government. It became a criminal offense for anyone to suggest that there could be a credible alternative to SAP.[10]

SAP and the crisis of democratic transition

When Babangida came to power in August 1985, he knew that 15 years of military rule had not reconciled Nigerians to the idea of permanent military rule. Early in the life of the regime, the government announced its commitment to the defense of human rights and to democracy. In January 1986 the ban placed on the National Union of University Students and the Nigerian Medical Association by the last military regime was lifted. That same month the Political Bureau was inaugurated. Its mandate was to carry out a nationwide debate and make recommendations on the political future of the country. Fifteen months later the Bureau submitted its report. Among other things it recommended the adoption of the American presidential system, the creation of more states and the initiation of programs to inculcate democratic and patriotic norms in the populace. As a follow-up to the Bureau's report, in July 1987 the Directorate of Mass Mobilization for Social Justice, Self-Reliance and Economic Recovery (MAMSER) was launched to educate, mobilize and prepare the electorate for the new dispensation. In March 1988, in addition to the National Population Commission, the National Revenue Mobilization Commission, the Code of Conduct Bureau and the Code of Conduct Tribunal were established.

In May 1988 the Constituent Assembly was inaugurated. A year later a new constitution was passed into law, the National Electoral Commission (NEC) was established, and the ban on politics was lifted. A multiplicity of parties came into being. In October 1989 the NEC recommended the recognition of six. The government, however, rejected the NEC's recommendations and banned all the parties. It claimed that none of the parties was able to meet the requirements that they not be regionally or ethnically based, but that they should reflect the federal character of the country and be solvent enough as not to be hijacked by any "money bag" (a euphemism for rich and discredited politicians). Instead, the government established two parties, both of which were financed by the government. The government also defined their ideological positions, with one being a little to the right and the other a little to the left. The two parties were the National Republican Convention and the Social Democratic Party, respectively. Discredited old politicians were banned from participating in politics while the new politicians were directed to find their company in any of the two parties.

The date for the transfer of power to civilians was shifted from October 1990 to January 1993, and in spite of mounting skepticism on the part of the intelligentsia and some politicians, the transition program took off well. Local government elections were held successfully throughout the country. In January 1992 elected civilian governors and members of the state assemblies were sworn in in all the 30 states of the federation. Similarly, the National Assembly elections were successfully held in July 1992. The elected members were, however, not sworn in until early 1993, largely because the government could not figure out in time what powers it was prepared to surrender to them and which to retain.

The presidential election, however, caused the most uneasiness. Twice in mid-1992 the presidential primaries were suspended and, as if to further confuse matters, the ban placed on discredited politicians was lifted. This changed the political equations overnight. The discredited politicians with their enormous wealth hijacked the political process and became the dominant contenders for the presidency. Eventually the government permitted the presidential primaries to be held in October 1992. The president, however, cancelled the results of the primaries, alleging widespread corruption and the use of money to buy votes. The cancellation was received with widespread cynicism as leaders of opinions all over the nation criticized the government, accusing it of having a hidden agenda. As rumors filtered around that the government intended to extend the transition to 1995, eminent Nigerians, such as General Obasanjo, a former Head of State, and others warned the administration that it would be insulting the nation's intelligence and courting disaster if it extended its stay in office beyond 1993. In this atmosphere of apprehension the government announced that it was postponing civilian rule until 27 August 1993, the date of its eight-year anniversary in office. Public cynicism and distrust reached such an extent that, as Obasanjo aptly put it: "when the government says good morning, people will look out four times to ascertain the time of the day before they reply."[11]

Amidst doubts and apprehension over the administration's commitment to handing over power, new primaries were held, culminating in the presidential election on 12 June 1993. As the results of the election began to come in, and it became evident that MKO Abiola, a multi-millionaire and the candidate of the Social Democratic Party, was set to win by a wide margin, the government stepped in. It stopped the announcement of the results, eventually annulling the whole election and suspending in its entirety the whole transition program on 23 June. The uproar and the sustained massive protest that greeted the cancellation took the government by surprise. It had underestimated the strength of the people's opposition and the level of the government's unpopularity. Unable to suppress the riot and squelch the mounting protest, Babangida was harried and hurried out of office in disgrace barely two months after the cancellation of the election results. An interim government, dominated by members of the Babangida's administration, took over. It, however, could not last as opposition continued. The helplessness and indeed powerlessness of the government to force its acceptance on the people only exacerbated the crisis. Eventually, in the midst of the continued protest and lawlessness that followed the ill-timed and ill-advised removal of the subsidies on petroleum products, the military once again took over power in November 1993.

Transition, popular culture and the re-awakening of civil society

In its attempt to win public support the Babangida administration went to great lengths to extend its influence over a wide spectrum of the various interest groups in the country. To ensure the loyalty of the army, the real basis of the government's legitimacy and power, steps were taken to eliminate all possible sources of disenchantment. Many serving and retired army officers were given shares in the government and other forms of inducement. Traditional rulers also came under increased government influence. In spite of modernization and the introduction of Western forms of government, especially at the local government level, traditional rulers have continued to exercise major influence in national affairs. Recognizing the continuing resilience of these traditional institutions, successive governments have over the years tried to enlist their support in the pursuit of their programs. Traditional rulers were regularly and routinely courted. They have been used as "firefighters" in times of serious domestic and national crises. In July 1993, for instance, a panel headed by the Sultan of Sokoto was largely instrumental in defusing the year-long protracted crisis between university teachers and the federal government. Babangida himself received many notable traditional titles from different parts of the nation.

The attempts of the administration to exploit the influence of traditional rulers sometimes resulted in troubles for the various parties involved. It is widely believed that the government's decision to secretly enlist Nigeria as a full member of the Organization of Islamic Conference was at the instance of the Sultan of Sokoto, the most prominent traditional ruler in the country. Similarly, after the cancellation of the 12 June election, the administration also tried to use the traditional rulers to appeal to the people and gain time as it temporized. The key traditional rulers were invited to Abuja, where the government attempted to win their support. After the meeting, Oba Okunade Sijuade, the Ooni of Ife and the most influential Yoruba ruler, was chosen as the spokesman for the royal fathers. His "double talk" to the press after the meeting, which endorsed Babangida's annulment of the elections, created problems for the monarch among his own people. He was denounced for allowing himself to be used as a stooge by the administration.[12] In the same vein, the Sultan of Sokoto was accused of complicity in the cancellation of the result of the elections. His failure to condemn Babangida's action, until it was too late, was used by his critics as a clear indication of his complicity. Similarly, the administration also sought the support of religious leaders, from both ends of the spectrum, through donations of large sums of money for the building of mosques and churches, and the granting of other favors.

Pro-democracy organizations and the struggle for democracy

Civil society was also affected in various ways by the struggle for democracy during the Babangida years. As the regime became more authoritarian and more intolerant of opposition, its intentions to hold onto power indefinitely became more evident. Consequently, concerned individuals and groups banded together to form new associations committed to defending human rights and democracy. Only the federal or national government can authorize the establishment of political parties, but other civil, religious, social and pro-democracy associations can be established with minimal cost and with very few bureaucratic controls.

The Babangida administration, through propaganda and the offering of incentives and grants, encouraged the establishment of civil and community-based organizations. It did this for two reasons. One was to give the regime an alternative base of support in order to compensate for the loss of support of the middle class, already eroded by the pangs of SAP. The other was to emphasize to the populace that the dwindling resources were making the government vacate much of the development space it had hitherto monopolized. The fall-out of this was the proliferation of trade, professional, youth, civil and pro-democracy associations, as well as women's groups. The government, of course, strove, though often without success, to deploy all the resources at its disposal to divide, coopt, subvert and remold the forces of civil society in order to serve its interests.

As early as 1981 the Committee of Concerned Citizens was set up to monitor and consider issues connected with the development of democratic tradition in the country. But not much could be achieved as military rule returned in 1983. As Babangida's strangulation of critical opinions intensified, and as his administration repeatedly reneged on its promise to re-establish democracy, the need to awaken Nigerian civil society to its responsibilities and its strength led to the formation of new organizations. In October 1987 two young lawyers, Olisa Agbakoba and Clement Nwanko, established the Civil Liberty Organization (CLO).[13] Its objective was to monitor and advance the cause of human rights in the country. Its immediate attention was directed toward exposing public and international attention to the inhuman conditions prevailing in many Nigerian prisons. By providing legal assistance, and by active collaboration with the press, it succeeded in securing the release of many prisoners and other detainees who were being held illegally or without trial. The CLO soon became popular, establishing branches in different parts of the nation. It became the ombudsman for individuals with an ax to grind with the administration. It furthermore began to prepare and publish for general circulation an annual report on the human rights situation in the country. The CLO documented and publicized many cases of the government's abuse of human rights, such as the arrest and imprisonment without charge of Gani Fawehinmi in a remote prison in the northern desert town of Gasua, popularly referred to as "Nigeria's Siberia" for four months. Fawehinmi had sought to prosecute two high-ranking state security officials for the murder of the *Newswatch* editor, Dele Giwa.

Before long it became apparent that the military could no longer be trusted to voluntarily return the country to democracy. Other legitimate means of opposition, such as the judiciary, the legislature, and the labor unions, were either silenced, neutralized, intimidated to submission or "settled."[14] The constant doctoring of the transition program also belied government's commitment to democracy. It became clear that democracy would have to be fought for, through a vigorous and systematic mobilization of the population. Thus, as the threats to democracy increased, new associations were established to push forward the struggle for freedom from corruption and authoritarian rule.

In February 1990 Nwanko and others formed the Constitutional Rights Projects (CRP). The group that was to have the most profound influence during the 1993 crisis was the Campaign for Democracy (CD). Gani Fawehinmi, the militant human rights lawyer; Beko Ransome-Kuti, the crusading one-time leader of Nigerian Medical Association; Olu Onagoruwa, a constitutional lawyer; as well as Chris Ubani and others, established the CD. One other group that emerged during the 1993 crisis was the Movement for National Reformation established by Anthony Enahoro, a former federal minister. Another was the high-profile Association for Democracy, Good Governance in Nigeria (ADGGN) established by General Obasanjo and made up of former heads of states and other prominent citizens. The caliber of its membership, as well as Obasanjo's courageous and uncompromising leadership, made the association a thorn in the flesh of the administration. As it turned out, Obasanjo was preparing the way for his own internment.

More importantly, however, the CD and the CLO became the rallying points of popular opposition to Babangida's attempt to subvert the democratic transition. Through press releases, pamphlets, handbills, public rallies and demonstrations, they provoked, organized and sustained popular opposition to the Babangida's regime until the very end. The arrest and detention of their leaders did not in any way diminish their effectiveness. So influential and so effective were these human rights groups that they succeeded in literally paralyzing life and activities in the southwestern parts of the country, and in virtually bringing the nation to a halt. They routinely exposed the selfishness and inordinate ambition of the military and their minions to public ridicule. They succeeded in mobilizing the public to actively and vigorously reject the "treacherous" attempts of the political class to take the nation for a ride and fool the public. That the campaign to remove Babangida was sustained and ultimately succeeded, even after the political class had virtually capitulated, was principally due to the resilience and courage of these groups and their allies in the media.

The prominence of the pro-democracy groups was due to a number of factors. One was the failure of the political class to remain steadfast in its opposition to military rule, its lack of genuine commitment to democracy, which would have required political and financial accountability to the electorate. The political class had sufficiently discredited itself by constantly playing into the hand of the military, as it splintered into self-seeking groups scrambling after immediate financial rewards and jockeying for positions. As inordinate power seekers, they often excelled the military in their engagement in gross electoral misconduct and corruption.

Another factor was the systematic emasculation of the judiciary which, having lost any semblance of independence, had become a veritable instrument of oppression to validate and legitimize any act of the military. In any case, it had been rendered impotent with virtually all its jurisdictions ousted by series of authoritarian decrees that placed the military and their acts above every law of the land. The opposition was further handicapped by the government's castration of the labor unions. Through the manipulation of the electoral process, and through the use of co-optation and financial settlement, the government had planted its agents in the leadership of the major labor unions and trade associations. Thus, by the time of the election crisis of 1993, the president of the central labor organization, the Nigerian Labor Congress, Paschal Bafyau, was generally derided by the workers for his crash opportunism, indecisive leadership and his brazen collaboration with the government. In 1994 Frank Kokori, the leader of the striking oil workers, challenged Bafyau to lead the union to be on the side of the people in the struggle against military tyranny. Bafyau simply retorted that his own survival was paramount and that he did not care on which side of history he was.

Even the professional bodies were not spared. The usually outspoken Nigerian Bar Association reneged in its opposition to the government, following the controversial election to its leadership of a government-sponsored candidate, Charles Idehen, to replace the more radical Alao Aka-Bashorun in 1989.[15] Only the Academic Staff Union of Nigerian University (ASUU) and the Nigerian Medical Association (NMA) were able to spurn government inducement and divisive tactics. For this the ASUU was summarily proscribed by the government in July 1988. The NMA was also crushed. Its frail but outspoken president and human rights activist, Beko Ramsome-Kuti, spent much of Abacha's reign in solitary confinement.

The strangulation and co-optation of these traditional opposition groups created a vacuum that the pro-democracy groups came to fill. They assumed the leadership of the opposition and became the mobilizers of civil society in the resistance against military authoritarianism. In an atmosphere of oppressive and exploitative military dictatorship, with the party systems officially suppressed or compromised, civil society could no longer limit itself to the more neutral and more primary function of deepening civic activities. It became a "refuge of last resort for partisan oppositional groupings."[16] The combination of the social trauma and economic misery of SAP, such as steep inflation, unemployment, malnutrition, deteriorating health care, armed robbery and violence and a high mortality rate, was a mixed blessing for the pro-democracy groups. All these created a general disenchantment with the government, while at the same time making the people more favorably disposed toward supporting the opposition and protest forces led by the pro-democracy groups.

The media and the transition crisis

One of the institutions significantly affected by the whole transition crisis was the press. The Babangida era, with all its active politicization of every aspect of the people's lives from money, to education, to religion, was a time of rapid expansion for the press, most especially the private press. Between 1985 and 1993 the number of private newspapers and weekly magazines doubled. From the inception of the administration, Babangida portrayed himself as a military democrat. Recognizing the influence of the press in affecting public opinion and establishing legitimacy for any government, the administration took steps to win the support of the press. It repealed the anti-press Decree No. 4 of the last military regime. Incarcerated editors and critics were released from detention. To seal the halcyon relationship with the press, Babangida appointed Duro Onabule, one of the most outspoken journalists, who was also the editor of one of the most critical private newspapers, *National Concord,* as his press secretary. Journalists were routinely invited to the State House and regularly feted and courted. Some of the leading journalists became unofficial consultants to the government as their views were sought with regard to issues of national and domestic interests. The Nigerian free press had never had it so good.

It is not surprising that many of the newspapers and magazines and their journalists became willing and enthusiastic supporters of the administration. The appointment of notable and respected human rights advocates, such as Professor Olikoye Ransome-Kuti, a member of the celebrated Funmilayo Anikulapo Kuti family, as the Minister of Health, and of Bola Ajibola, the controversial and outspoken critic of the previous administration and President of the Nigerian Bar Association, as Minister of Justice and Attorney General, gave the administration a populist image and seemed to lend practical proof to its avowed commitment to the protection of human rights and democracy. This image was later reinforced by the appointments of Dr Tai Solarin, a noted social critic, as the chairman of the People's Bank and of Professor Wole Soyinka, the Nobel Laureate (1986) and playwright, as the chief of the newly established Federal Road Safety Corps (FRSC).

The honeymoon, however, did not last for too long. As the administration tightened its hold on power, and as the true character and intentions of the president became clearer, it became increasingly evident that the self-proclaimed military democrat had little commitment to honoring his pledge to hand over power to an elected government. One by one the private newspapers began to extricate themselves from the favors and overtures of the administration. Before long they became rather vocal in their criticisms of the growing official corruption and the reckless mismanagement of the nation's resources. As the criticism became more strident and embarrassing, and the press became more militant, the government's initially favorable overtures to the press turned to hostility. The press became an enemy to be hunted down. By 1986 the innovative and trail-blazing newsmagazine, *Newswatch*, led the way in this unsparing assault on the administration. After several weeks of arrests and interrogation by the National Security Organization (NSO), the founding editor-in-chief of *Newswatch* and the most popular journalist of the time, Dele Giwa, was killed by a letter bomb widely believed to have been sent to his home by the government's secret service unit. Public support and sympathy for the harassed press, and the open implication of the administration in the murder, only succeeded in further emboldening the press in its self-avowed mission of being the people's watchdog and mouthpiece. The government, however, remained unrelenting in its determination to brook little opposition. In the last few years of the administration virtually all the private newspapers and magazines were repeatedly and routinely shut down and re-opened by the military. Journalists who refused to be intimidated, blackmailed or "settled" were regularly arrested and incarcerated.

This state of uneasiness reached its climax during the 1993 election crisis. With all other avenues of protest and resistance effectively muted and muzzled, and with leaders of the opinion "developing tight lips and withered hands,"[17] the press took on the responsibility of articulating and giving expression to the will of the people. The press also became the main vehicle for the dissemination of the ideas of the pro-democracy groups. In sensational and often provocative editorials and headlines, the press time and again took the government to task, casting doubts on the transition program and raising awkward questions concerning the government's intentions. It popularized and sensationalized evidence that belied official pronouncements and revealed the hidden agenda of the administration.

The cancellation of the 12 June election result was the final straw that shattered whatever moderation the press had exercised in its relation with the government. It confirmed the worst fears of the press, which came down heavily and unsparingly on the administration.

The government, on its part, responded in kind. It directed all federal government-owned media, such as the *Daily Times*, the *New Nigerian*, the National Television Authority, and other pro-Babangida media to launch a counteroffensive against all opponents of the government. In sponsored advertisements and publications in newspapers and in the electronic media, most of which it controlled, the administration contested the opposition press for the support of the people. This, however, did not help the government, as its popularity declined. The pro-government press suffered a drastic loss in readership as the opposition press was having a free day. Soon it became clear to all the parties involved that the contest for public support was being fought and won in the media. Unable to beat the media houses at their own game, the government took stronger measures. Two weekly magazines, *Tell* and *The News*, the most consistent and the most popular critics of the administration, were the first to fall under the government sledgehammer. Intensified harassment led to the seizure of well over 300 000 copies of the magazines from their printers.

To elude government agents the magazines changed their format to tabloid that could easily be printed in black and white in any of the numerous presses without detection and seizure. Eventually the two newspapers were proscribed and their editors declared wanted for treasonable felony. In response *The News'* editors went underground and re-emerged with a new tabloid, *Tempo*. *Tempo* was an instant success. It became the most popular and most widely read tabloid in the country. All attempts made by the regime to find where the illegal newspaper was being printed, and to track down its elusive and "ghost" editors, proved abortive. Exasperated, the government in desperation sent out soldiers to patrol the streets, seize the subversive papers from vendors all over the country and arrest anyone caught reading the offensive publications. As international opposition to the government mounted, and various sanctions began to be imposed by western nations, the Secretary of Information had to call a world press conference to announce that the administration had sufficient evidence to prove that

Tempo was being printed inside the American Embassy. This was a diversionary tactic. It did not work.

As unabating demonstrations and brutal repression, especially in the southwestern part of the country, rocked the nation, the government came under a chorus of criticisms. Prominent citizens, including former heads of states, called on the government to reverse its decision, accept the verdict of the 12 June election and disengage the army totally from politics. Feeling boxed into a corner, and infuriated at the unrelenting press attacks, the government took more desperate measures. On 22 July, in an unprecedented move and in one fell swoop, it shut down five newspapers and one radio station. The most hard-hit was the Concord group, comprising about seven publications and owned by Abiola, the winner of the 12 June election. The others were *Punch*, the *Abuja Newsday*, and the *Sketch*, owned jointly by the governments of the Oyo, Ogun, Ondo and Osun states, whose governors had left no-one in doubt as to their support for Abiola and their opposition to Babangida. The *Reporter*, shut down the previous year, remained shut. Much pressure led to the re-opening of the Ogun State Broadcasting Corporation, but the *Observer* newspaper, owned by the Edo state government, took its place. New decrees prescribing draconian punishments for erring newspapers and journalists formally legalized what the government was already doing in action. Closure orders were enforced through round-the-clock sieges of offending media houses.

Popular culture

The successful clampdown on, and the military siege of, the critical media, however, did not mean the end of the opposition to the administration. The remaining newspapers on the two sides of the struggle, and those who claimed to be neutral, continued in various ways to express their views of the administration. The use of cartoons became an effective means of expressing opinions about the administration. Day after day the public was bombarded with different kinds of cartoons implicitly cynical and covertly critical of the government. These cartoons took on a life of their own. Unlike before the election, they were no longer generally limited to the media. They began to appear in locally produced books, pamphlets, handbills and leaflets. These leaflets became popular and were widely photocopied and circulated freely among the population. Many of them were produced in or translated into many of the indigenous languages. Produced anonymously, and often by pro-democracy groups, such as the Campaign for Democracy and the Civil Liberty Union, or under false or pen names, they became means of laughter in the midst of tension. They also served to reinforce and meet the people's need for means of voicing their opposition to the administration without running afoul of the law. In their expression they were usually very graphic, vivid, funny, playful and imaginative.

The political crisis became a catalyst for innovative arts. The protagonists in the conflict, and most especially the president, were given various connotations. Babangida got many new names, titles and aliases.

It is also interesting to note much of the artwork, especially the cartoons produced outside the media houses, was not carried out by professional artists but by amateur or popular artists who were responding in their own way to the unfolding crisis around them, and thereby helping to articulate public attitude and shape public opinion. While some of them exploited the opportunity to make money and practice their art, they had no doubt that they were engaged in a major sociopolitical discourse of national significance. But it was not only the president who became a victim of the new aggressiveness of popular art in this period. There were also many cartoons ridiculing and castigating the politicians for their cowardice, sycophancy, hypocrisy, proverbial susceptibility to manipulation and corruption, as well as their betrayal of the people.

An epilogue: Abacha's dictatorship, 1993–98

Five years after the abrogation of the 1993 election result, Nigeria remained in deep crisis. General Sanni Abacha, who came to power in the wake of the election imbroglio, unleashed a reign of terror on the opposition. This was meant to purge out the memory of the 12 June 1993 election and pave the way for Abacha's self-succession and the perpetuation of military rule. Having tasted the pie of politics and the spoils of power, the army had no intention of again relinquishing power. As a former military Head of State of Nigeria declared in June 1998, "Our military personnel have generally become inured to corruption, lying, selfishness, lack of patriotism, avarice and character and behavior unbecoming of a good military."[18] When Abiola, the winner of the 1993 election, declared himself president in 1994, Abacha promptly clamped him in jail. Challenged by the organized labor union of petroleum workers, who brought the nation to a halt for about two months in late 1994, Abacha quivered just a little, but soon regained control. The workers' strike was brutally crushed and its leaders, Frank Kokori and Milton Dabibi, tossed into the dictator's dungeon. Thereafter, the government descended on other critics, politicians, journalists, lecturers, students and union leaders, blackmailing, intimidating, and harrying them to jail. Most were kept in solitary confinement, denied visitation rights and medical treatment, and slept on the bare floor in some of the most unhealthy, overcrowded and poorly kept prisons in the world.

No-one was sacrosanct. Everyone came under Abacha's sledgehammer. Among those who succumbed to diseases in prison was General Shehu Yaradua, the one-time deputy to the former Head of State, General Olusegun Obasanjo. Obasanjo was also thrown in jail for his prominence in opposing Abacha's government. Another was Alfred Rewane, the 79-year-old leader of the National Democratic Coalition (NADECO, a political pressure group established in April 1994). In October 1995 three well-dressed men suspected to be agents of the government gunned down the octogenarian nationalist and philanthropist in his living room.[19] Even traditional rulers were not exempt. Ibrahim Dashiki, the Sultan of Sokoto and the most prominent natural ruler in the country, came under the lash. Suspected of wavering in his support of Abacha's dictatorship, he was summarily deposed and sent into internal exile. Even those in exile, who have not kept quiet and made the fatal mistake of daring to visit the country, walked into the dictator's trap.

Human rights and minority group rights advocates, such as the writer, Ken Saro-Wiwa, were summarily tried and sent to the gallows. Independent and critical media houses were besieged by heavily armed soldiers, who seized their workers, impounded and set ablaze freshly produced newspapers and magazines, vandalized and ripped off computers and printing equipment from the walls. Hired assassins prowled the country gunning down the most inveterate of Abacha's enemies, and at times their wives and children.

Bespectacled behind dark glasses, taciturn, inscrutable and unforgiving, Abacha had the distinction of being the most brutal military tyrant in Nigerian history. A former Head of State, and one of the many to feel the brunt of his whip, in a sentiment shared by most of his countrymen, described Abacha's rule as the most "deceitful, Godless, corrupt, soulless, oppressive, murderous, obstinate and wicked regime" in Nigeria's history.[20] Obsessed with security, recluse, rarely seen except by his coterie of his hangers-on, and cacooned away in his highly garrisoned and inaccessible Aso Rock fortress, Abacha will go down in Nigerian history as the most inscrutable and enigmatic despot in the country's history. Nevertheless, all this notwithstanding, five years after the election crisis, thanks to the resilience and unflagging resistance of the Nigerian pro-democracy groups, 12 June has refused to go away. Though severely tried and chastened, Nigeria's pro-democracy groups have refused to surrender.

Some conclusions: a nation at the crossroads

Being present in Nigeria during the 1993 crisis, and having personally visited a few times since then, it is indeed tempting to exaggerate the achievements of the pro-democracy groups and other elements of the emerging civil society. By the time the crisis was temporarily defused with the forced exit of Babangida, the pro-democracy groups had become a force to reckon with in the game of power politics in the country. It is indeed instructive to note that when, in November 1993, senior army officers came together to plan the return of temporary military rule as a way out of the deepening political crisis, the issue that bothered them most was not the likely resistance of the politicians, but how to handle the popularity and effectiveness of the pro-democracy groups. Developments in Nigeria between 1993 and 1998 have shown the limit and indeed the futility and danger of coercion in the face of an awakened, politically sensitized and determined citizenship. Babangida's failure and his fall from office were widely seen as an indictment of the military, but the politicians did not fare better either. Their willingness to compromise every principle for money, and their proverbial readiness to be co-opted and "settled" gave them a negative image before the public. The only groups with considerable respect and credibility were the pro-democracy groups, whose leaders were widely regarded as heroes and champions of the people. Even the military oligarchy recognized that to ignore the "power" of the pro-democracy groups is to do so at its own peril. They kept the embattled security agents of Babangida and Abacha busy, and ensured that, for as long as military dictatorship persists, the enemies of democracy would also have no rest.

Thus, the contribution of the pro-democracy groups to the struggle for democracy in Nigeria should not be underestimated. Beyond their immediate achievements their significance as catalysts of democracy and as facilitators in the resurgence of civil society must be recognized. As independent and autonomous groups free from official control and funding, they provided the basis for the re-awakening of civil society in the face of authoritarianism. Through their activities, and especially their outspoken assertiveness, they helped to sustain democratic ideas and values. Through their resistance to dictatorship and oppression in all their ramifications they helped to prepare the way for the emergence of an articulate, vibrant and pluralistic civil society, without which a democratic tradition cannot develop. By organizing and encouraging the active participation of the majority of the population in politics, they rescued politics from being the exclusive preserve of a tiny minority. They made the Nigerian people become actively involved in the determination of their future and the destiny of their nation.

Abacha is dead: Nigeria managed to survive him. His Chief of Army Staff, General Abubakar, is now at the helm. The strange death of Abiola, ostensibly from an apparent heart attack on 7 July 1998, while still in government custody and on the day before he was to be released, was seen by many as being too convenient and suspicious. The news provoked widespread rioting and bloodletting in the Nigerian southwest and among the Yoruba. There were strident calls for the Yoruba to secede from the country or, failing that, for the nation to adopt a confederal system of government.

Open and free discussion of all issues, mutual trust and sincerity rather than suspicion, co-optations, threats and force is the way forward. Intimidation and sequestration of opponents and suppression of legitimate grievances or the customary practise of sweeping the problems under the carpet in the name of peace and national unity will only postpone the evil days. It is only against this background that the continuing convulsion over the 12 June election can be understood. That date should be seen as a catalyst triggering off deep-seated, though covered-up, but largely latent, antipathies.

It therefore came as no surprise that all three political parties chose to nominate only Yoruba as their presidential candidates. This was done to appease the Yoruba who had been collectively alienated over the 12 June 1992 annulment, and thus forestall the seemingly inexorable slide toward disintegration. The new civilian regime of General Obasanjo will have to confront and overcome many challenges if it is to go down in history, not as the last stage of a military transition, but as the first stage in Nigeria's transition to full democratic rule. This will not be an easy task. Having been largely bankrolled into power by a coalition of retired but still powerful generals led by Ibrahim Babangida and other rich politicians, noted for their thievery and mismanagement of Nigeria's enormous oil wealth, Obasanjo will have to break out of their stranglehold. For Nigeria to move forward in the direction of real democracy, and for the lives of its impoverished and battered millions to have any chance of future improvement, Obasanjo must be prepared to take drastic but necessary measures to check corruption, stop money laundering, and depoliticize and tame the army, consigning it to purely defensive roles. He must act decisively to reverse the poverty and ecological devastation of the oil-producing Delta communities, calm ethnic tensions, especially through a restructuring of the country that will shift power from the center to the component parts, and return the nation to true federalism.

Whatever lesson we can learn from the Nigerian experience, it is evident that in Africa's attempt to find a way out of the quagmire of its sociopolitical predicament, there is no alternative to the development of a democratic tradition, the empowering of the populace, the enthronement of responsible leadership and the entrenchment of freedom of expression. In the meantime, as the nation's economy continues to deteriorate and as the country teeters on the verge of explosion, it is clear from this chapter that the solution to the Nigerian crisis and the way forward lies in the re-awakening of civil society and its seizing the political and economic initiatives from the old ruling military and civilian aristocracies who had between them dominated, ruled, divided, politicized ethnicity and religion, looted and ruined the country with reckless abandon and with far-reaching consequences, since independence in 1960.

Notes

A shorter version of this chapter was first presented at the University of Witwatersrand History Workshop on "Democracy: Popular Precedents, Practice and Culture," Johannesburg, South Africa, July 1994.

1 Civil society in this chapter is defined as "the realm of organized social life that is voluntary, self-generating, [largely] self-supporting, autonomous from the state, and bound by a legal order or set of shared rules ... it involves citizens acting collectively in a public sphere ...[as] an intermediary entity, standing between the private sphere and the state." Larry Diamond, "Rethinking Civil Society: Toward Democratic Consolidation," *Journal of Democracy*, 5, 3 (July 1994), 5. See also N. Chazan, "Africa's Democratic Challenge: Strengthening Civil Society and the State," *World Policy Journal*, 9 (Spring 1992), 279–308. On popular culture see C. Mukerji and M. Schudson (eds), *Rethinking Popular Culture: Contemporary Perspectives in Cultural Studies* (Berkeley, CA: University of California Press, 1991),

pp. 1–55, and the interesting introductory essay by Karin Barber, "Popular Arts in Africa," *Africa*, **30**, 3 (September 1987), 1–78.

2 For more on the historical origins and development of Nigeria as a political entity see Funso Afolayan, "Nigeria: a Political Entity and a Society," in Paul A. Beckett and Crawford Young (eds), *Dilemmas of Democracy in Nigeria* (Rochester, NY: University of Rochester Press, 1997), pp. 45–64.

3 On the failure of Nigeria's First Republic see R. L. Sklar, *Nigerian Political Parties: Power in an Emergent Nation* (Princeton, NY: Princeton University Press, 1963) and L. Diamond, *Ethnicity and Democracy in Nigeria* (New York: Syracuse University Press, 1988).

4 On the collapse of the Second Republic see T. Falola and J. Ihonvbere, *The Rise and Fall of Nigeria's Second Republic, 1979–1983* (London: Zed Books, 1985) and R. A. Joseph, *Democracy and Prebendal Politics in Nigeria: The Rise and Fall of the Second Republic* (Cambridge: Cambridge University Press, 1987).

5 F. J. Ramsay, *Africa: Global Studies* (Guilford, CT: Duskin/McGraw-Hill, 1993, 1997), pp. 54–7. See also World Bank, *Nigeria Structural Adjustment Program: Policies, Implementation, and Impact* (Washington, DC: World Bank, 1994).

6 Thomas Adejo, "13,767 Nigerians Seek for Asylum in Germany," *National Concord*, **16** (June 1992).

7 For these statistics from the Federal Ministry of Health, Lagos, 1990, see Deji Popoola, "Nigeria: Consequences for Health," in Aderanti Adepoju (ed.), *The Impact of Structural Adjustment on the Population of Africa* (Portsmouth, NH: Heinemann, for the United Nations Population Fund, 1993), pp. 93–6.

8 For instance, one of the 1989 riots was provoked by reports in some Nigerian newspapers that a United States-based magazine, *Ebony*, had named Babangida as one of the richest African rulers. Fifty-five people died in the riot. See Larry Diamond, "Nigeria's Third Quest for Democracy," *Current History*, **90** (May 1991), 203.

9 Julius Ihonvbere, "A Critical Evaluation of the 1990 Failed Coup in Nigeria," *Journal of Modern African Studies*, **29**, 4 (December) 1991, 601–26.

10 "SAP: What Alternatives," *Newswatch*, Lagos, 24 July 1989, pp. 14–15; Soji Akinrinade, "Is SAP the Magic Wand?," *Newswatch*, Lagos, 5 October 1989, and "Where do We Go from Here?," *Tell*, 7 October 1991.

11 *Tell*, 26 April 1993, p. 21. For an explanation of the election crisis and the problem of ethnicity see C. S. Whitaker, The *Politics of Tradition: Continuity and Change in Northern Nigeria, 1946–66* (Princeton, NJ: Princeton University Press, 1970).

12 "How Not To Talk," *Tempo*, 26 July 1993.

13 On the establishment of the Civil Liberty Organization see C. Nwanko, "The Civil Liberty Organization and the Struggle for Human Rights and Democracy in Nigeria," in L. Diamond, *The Democratic Revolution: Struggles for Freedom and Pluralism in the Developing World* (New York: Freedom House, 1992), pp. 105–23.

14 This was a euphemism for being bribed with money, contract awards, lucrative appointments, or allocation of amenities to one's community, to name a few commonly used inducements during these years.

15 "NBA: The Change of Batton," *Daily Times*, Lagos, 15 September 1989.
16 M. Bratton, "Civil Society and Political Transition in Africa," in J. W. Herbeson, D. Rothchild and N. Chazan (eds), *Civil Society and the State in Africa* (Boulder, CO: Lynne Rienner, 1994), p. 64.
17 Obansanjo, *Tell*, 7 June 1993.
18 "Obasanjo Speaks on Coup Trial," *Sunday Vanguard*, Lagos, 21 June 1998.
19 "Why NADECO Chief was Murdered," *Tell*, 23 October 1995.
20 "Obasanjo Speaks on Coup Trial," *Sunday Vanguard*, Lagos, 21 June 1998.

Part III

The Middle East and North Africa

7

The Economic and Political Impact of Economic Restructuring on NGO–State Relations in Egypt

Janine A. Clark

Introduction

The new economic model of economic restructuring, deregulation and privatization, as implemented by international bodies such as the IMF and World Bank, is theoretically expected to lead to greater political liberalization and ultimately democratization in the countries of the Third World. While the process is not necessarily expected to be a smooth one, in the long term the withdrawal of the state in the economic sphere, and consequently its reduced role, is assumed to provide both economic and political opportunities for non-state actors. In simplistic terms, the greater economic role of the private sector should ultimately lead to a greater political role in policy- and decision-making.

This chapter examines the degree to which this process is happening in terms of the non-governmental sector in Egypt. In particular the chapter focuses on non-governmental organizations (NGOs) dealing in the field of social welfare (as opposed to environment, advocacy or charity work, for example) and their changing relationship with the state.

The impact of economic restructuring and privatization upon social-welfare NGOs and their ability to successfully "step in" and replace the state as it withdraws involves two interrelated issues. The first is the service capacity and sustainability of the NGOs – their economic status, their level of expertise, their management and organizational skills. The second is political: whether they will have the political "space" within which to flourish and influence policy- and decision-making.

This chapter argues that, in the face of rising inequalities, social-welfare NGOs have neither the service capacity nor the sustainability to address the needs of the increasing numbers of poor. On the one hand, the majority of NGOs are too small, offering very few services, and suffering from a lack of trained staff and organizational skills. On the other hand, a large percentage of the small number of very successful NGOs are highly dependent on state and foreign aid, thus rendering their sustainability also into question. More importantly, the NGO sector has not been given any political space within which to increase its political role in terms of the lifting of any state restrictions and regulations. There is, furthermore, evidence that, as privatization proceeds, ministerial attempts to control the sector may intensify. These political ties and regulations which the state exerts over NGOs' continue to limit NGOs' abilities to address their organizational and financial weaknesses and to play any role in policy-making. Economically and politically, Egypt's NGOs are ill-equipped to deal with the increasing numbers of people negatively affected by the government's economic reform program or to manage a large and complex field such as social welfare.

Corporatism in Egypt

While Egypt is a semi-corporatist state, it is within the framework of corporatism that state–civil society relations must be viewed.[1] Corporatism reached its zenith under the leadership of Egypt's first president, Jamal Abdel Nasser (1952–70), and continues to form an effective means of state control today. Within the context of his larger program of socialism and nationalization, Nasser abolished all political parties, created a one-political-party state, and ultimately restructured group activity in each occupational sector into new corporatist groups producing a nearly universal system of functional representation under state control. Today, representation in each occupational category, whether it be doctors, lawyers, or workers, remains limited to a single monopolistic association to which membership is mandatory for all who work in its field of jurisdiction. Organizationally the structure of the syndicates and unions is hierarchical with regional branches and, finally, national confederations that are linked to their respective parent ministries. The corporatist system thus is an effective means of top-down control of the syndicates.

The concentration of power in the state's, and specifically the president's, hands continued under president Anwar Sadat (1970–81), and was reflected in the 1971 Constitution (still in effect today). The Constitution delegates the president of Egypt as the head of state, supreme commander of the armed forces, as well as the head of the executive. Both the prime minister and the legislature are also responsible to him. As such, the executive branch has the power to act on any significant issue even without the support of the legislature.

The president's power is further enhanced by his dual role as president and as head of the majority party, the National Democratic Party (NDP). This means that, in essence, the opposition in parliament cannot block the government on any issue or bring it down.

Upon first coming to power Sadat made a few tentative steps towards pluralism. By 1978, for example, he had allowed six parties to establish. Eventually, however, he became unwilling to tolerate the criticisms that his limited liberalization permitted and, as tensions rose within the country, Sadat amended the constitution, and further curtailed the freedom of the press and the right to form political parties.

Sadat, moreover, reshaped the existing set of corporatist associations into more reliable instruments of control and cooptation. Sadat blatantly interfered in syndicate elections by rounding up busloads of his supporters to vote for the government candidate.[2] In a gesture towards appeasing the opposition, he created an upper house of parliament, half of which was composed of appointed representatives from occupational groups and national institutions. However, it was granted no real powers.

Since coming to power after Sadat's assassination in 1981, Hosni Mubarak has further reinforced the exclusion of political opposition from any real political power or influence. Mubarak has not relinquished state (or NDP) control over all aspects of political parties. Foremost amongst these measures are the emergency laws, first implemented as a result of Sadat's assassination, which are still in effect and further concentrate power into the president's hands. In addition, in 1983, new electoral laws were passed that inhibit any smaller and/or new parties from gaining access to parliament, and virtually ensure a majority of two-thirds of the votes for the NDP.[3]

Certainly, in his bid for national reconciliation and legitimacy after the violence and tension surrounding Sadat's death, Mubarak institutionalized greater steps toward political liberalization than his predecessors. Most important among these were the release of political prisoners, the restoration of freedom of the press and the renewed activities of political parties. Since the onset of Islamic violence in the 1990s, however, Mubarak has reviewed these policies and has embarked upon a dramatic shift in his approach to Islamic groups, including Islamic NGOs, and civil society in general.

Egypt's economic reform and structural adjustment program

In an effort to alleviate its US$40 billion debt, Egypt agreed to an Economic Reform and Structural Adjustment Program (ERSAP) with the IMF and World Bank in 1991.[4] The agreement with the IMF paved the way for a debt-reduction deal amounting to US$3 billion, to be completed in three stages, with the Paris Club.[5] Egypt has thus embarked upon an economic reform program based on macroeconomic stabilization and the structural transformation of the economy. The aim of the program is to reorient Egypt's state-directed economy toward market forces and the private sector, and external trade.[6]

Due to its reliance on state expansion, import substitution and to a large degree the oil shock, Egypt has been experiencing serious external debts since the mid-1970s. However, until the 1980s it had been more or less able to finance these deficits and ward off structural adjustment reforms through external rents. By the mid-1980s these rents had dried up: foreign worker remittances fell; oil revenues dropped; and Suez Canal receipts and tourism substantially decreased.[7] To aggravate matters, aid agencies from the IMF to unilateral lenders, such as the United States, began to more seriously link aid to progress in substantive economic reform, including privatization.

The case for privatization in Egypt rests on two bases: the size of the public sector and its performance.[8] By the mid-1980s about two-thirds of the non-agriculturalist workforce in Egypt was on the state payroll and these figures do not include the 400 000-member armed forces or the 150 000-strong police force.[9] The mammoth and inefficient public sector accounted for 90 percent of Egypt's foreign debt.[10] Of the total public sector debt, a conservative estimate put the share of public "profit-making" enterprises (excluding financial and defense sectors) at about one-third.[11] The public sector in Egypt has hence been a major contributor to Egypt's slowdown of economic growth, its poor export performance, heavy external debt, widening food gap, and its unhealthy dependence on foreign assistance.[12]

As a result of the failure of state economic enterprises and state intervention, Sadat introduced, and recently Mubarak is more forcefully pursuing, economic liberalization. Under IMF pressure, Mubarak has reluctantly undertaken incremental economic reforms toward privatization and the reduction of subsidies. Given the heavy toll these measures are taking on the public in terms of increased unemployment and rising costs of living, it is not surprising that the regime has hesitated.

Mubarak and his government have not forgotten the 1977 food riots in reaction against the removal of food subsidies by Sadat.

In 1987 Mubarak reached a agreement with the IMF for a loan of US$342 million. After the IMF dispensed the first installment of this loan (US$160 million) in 1988, however, the agreement collapsed. The IMF accused the Egyptian government of reneging on the necessary economic reforms; Egypt claimed it was merely slowing down the pace of reform for fear of massive social unrest.[13] Periodic talks after 1988 finally culminated in a new agreement in 1991. The Letter of Intent, signed by both Egypt and the IMF, carefully stipulated Egypt's commitment to specific economic reforms and liberalization.

There has been economic improvement under the present economic restructuring program. In terms of macroeconomic stabilization, the correction of price distortions, financial liberalization, and the removal of trade barriers, for example, Egypt has achieved remarkably positive results. Since the change in government due to the November 1995 elections, privatization of the public sector and companies is also being tackled more seriously. Due to this economic restructuring program and the huge infusion of aid after the Gulf War, Egypt has recently experienced economic success at the macro-level. Government spending has decreased, its revenue has increased, the budget deficit has decreased, the external debt has decreased, and inflation rates have lowered.[14]

For the poor in Egypt, however, unemployment and poverty are growing. The dramatic improvement at the macro-level is not found at the micro-level; In fact, quite the opposite is true. Micro-conditions have never been so bad: wages and salaries have fallen; average percapita yearly incomes have fallen; prices on foodstuffs, transportation, and rents, gasoline and electricity have risen as subsidies have been removed; the incidence of malnutrition has risen (after steadily falling from the early 1950s to the late 1980s); and economic growth has slowed.[15] According to the Egyptian Planning Institute the rate of poverty in Egypt reached 48 percent of the total population in 1995–96. Thirty-three percent of the population is living in absolute poverty. The same report notes that unemployment has reached more than 17.5 percent of the total manpower – 68 percent of which are high-school and university graduates generally under 25 years of age.[16]

Egyptian NGOs are already laboring under the increase in numbers of people seeking their services, yet the worst is still to come. The Egyptian government is presently under increasing pressure from the IMF to accelerate the privatization of its public sector as the next challenge in the economic reform program. This entails not only the sale of state-owned companies but also a decrease in state expenditures through, for example, the shifting of public service expenses to the private sector. The impact of these cuts, at least in the short run, is twofold. The first is a decrease in wages and an increase in unemployment through, for example, the recent elimination of the Graduates Program according to which university graduates received automatic employment in the state. This translates quite simply into poor and middle-class Egyptians having even less to spend.

The second is the need for Egyptians to find better social-welfare services elsewhere than the state. In this regard ERSAP is having its greatest impact in the area of health and education which traditionally have been heavily funded. In the area of health care, for example, figures:

> show a downward trend in the Ministry of Health's medi-care expenditures in relation to the State budget and the Gross Domestic Product (GDP): from 2.3% to 1.21% in 1981 to 1.9% and only 0.6% in 1993/94. Hence, government hospitals suffered a shortage in resources which led to neglecting renovations, maintenance and a sharp drop in medical facilities, causing widespread dissatisfaction with the offered services.[17]

As a result, the percentage of the population with access to health care has dropped since the early 1980s.[18] Statistics show that the population per physician increased in numbers from 732 in 1980–85 to 1316 in 1989–94. Similarly, the population per hospital bed grew from 483 to 517 during the same time period.[19]

Public expenditure on basic social services as a percentage of GDP has also dropped from 5.3 percent in 1980–85 to 4.9 percent in 1989–94.[20] For example, the number of new social units – government extension services that provide guidance in birth control, literacy programs, vocational training, cultural awareness, child welfare, and cooperative purchasing programs for raw materials and tools – built between 1988/89 and 1992/93 declined by 33.3 percent.[21] This lack of social units particularly affects Egypt's rural areas.

Egyptians are turning increasingly toward the private sector for their social-welfare needs. As the majority of Egyptians cannot afford services offered in private practises or companies, the bulk of this load has fallen on the shoulders of the NGO sector. NGOs as well as international donors are presently trying to deal with this increased role in as effective a manner as possible.

The diversity of Islamic activism in Egypt

Politically, economic reform has not been the only challenge confronting the Mubarak regime. Since the early 1990s Egypt has been the site of violent confrontations between Islamic political activists and the Egyptian government. The two – economic reform and Islamic violence – are not, however, entirely unrelated.

Broadly speaking, Islamic activism in Egypt can be categorized into two types: moderate and militant Islam. The moderate stream of Islamic activism comprises groups and associations consisting of the Muslim Brotherhood, its sympathizers, thousands of Islamic NGOs and numerous Islamic investment companies. Militant Islamic activism primarily includes three main militant Islamic groups – *Jihad*, the *Takfir wa Hijra*, and the *Jama'a Islamiyya* – and forms just a small part of the greater Islamic "movement." While both desire the establishment of an Islamic state, they differ significantly in their methods.

The Muslim Brotherhood is composed mainly of middle-class professionals – doctors, lawyers, journalists. While the Brotherhood is technically illegal and banned both as an organization and a political party, it participated in the 1984 elections in an alliance with the *Wafd* party and again in the 1987 elections in a coalition with the Socialist Labor Party and the Liberal Party. After having boycotted the 1990 elections, along with most other opposition parties, the Brotherhood once again allied itself with the Labor Party in the November 1995 elections.

Over the past ten years the Brotherhood has also run in most of the country's professional syndicate elections. Despite previous governments' interference in the elections and workings of the syndicates, under Mubarak the professional union elections have come to be considered, until recently, the only free elections in Egypt and have therefore become an important stronghold in the battle for greater democracy in Egypt. Since the mid-1980s Islamist political parties have repeatedly won elections in the doctors', engineers', pharmacists' and the lawyers' syndicates.

Ideologically influenced by, but not necessarily related to, the Brotherhood are the numerous Islamic social-welfare organizations or NGOs and economic enterprises. Often referred to as social Islam, this stream is engaged in setting up social-welfare services and economic institutions along "Islamic lines." This includes the establishment of Islamic banks, investment companies, and factories, as well as social and educational services such as Islamic medical clinics and welfare organizations. Two recent studies estimate Islamic social welfare organizations as the second-largest category of civil associations; the largest being government-created Community Development Associations.[22]

Economic reform and the rise of Islamic militancy

The roots of Islamic activism can be found in the social, economic, political, cultural, regional and international realms. Most analysts, however, point to the worsening economy and the government's neglect of the south as the primary catalysts of the recent violence in Egypt.[23] It is no coincidence, as Saad Eddin Ibrahim points out, that the most recent radical Islamic unrest began only shortly after the signing of the 1991 IMF deal with its consequent threat of job losses and cuts in the already inadequate government subsidies and services.[24]

Egypt's IMF economic reform program as implemented by the Mubarak regime thus far has not created the broad-based economic growth the economy needs to employ and feed the bulk of its population. As the gap between rich and poor is becoming greater, more people in Egypt, and especially in southern Egypt, are sharing less. It is in the north, and in Cairo in particular, that the government spends its money, builds its factories and institutes new social programs.

It is for this reason analysts agree that the violent and often deadly confrontations between Islamists and the security forces are primarily concentrated in the southern villages and provincial capitals, and in the urban fringe shanty towns of Cairo, Alexandria, Giza, Port Said and Suez in northern (Lower) Egypt. While violent slum areas are also located in the north, they are dominated by southerners who migrated in search of jobs. The socioeconomic profile of the militants produced in these rural areas, and urban shanty-towns and slums, reveals that they tend to be in their early 20s, lacking education and poor – those who feel they have nothing to lose and everything to gain by disrupting the present socioeconomic order and policies.[25]

The result is that the 1990s have been Egypt's bloodiest since the turn of the century. In the hopes of bringing down the Egyptian government by isolating Hosni Mubarak's regime internationally, and depriving it of its much-needed hard foreign currency, Islamic militants have extended their use of violence to include attacks on foreign tourists, Egyptian Christians, as well as numerous assassination attempts on public figures (including a number of attacks upon Mubarak's life). In addition to the hardships inflicted by the near state of war in certain areas of the country, the killings of foreign tourists by Islamic militants have succeeded in harming Egypt's tourist industry – the country's number one source of foreign exchange.

Islamic militancy and the government response

In an effort to crush the militants the government has adopted a more confrontational approach to Islamic activism than had been practiced during Mubarak's first decade in power. Upon coming to power in 1981 the Mubarak government tried to steer a course between suppression and a highly orchestrated campaign to make itself seem more Islamic. Aiming to curb the radical Islamic fringe without alienating the rest of the Muslim population, the government's strategy throughout the 1980s was to draw moderate elements, such as the Muslim Brotherhood, into its fold and marginalize the radical elements.

The change in the government's approach has meant that the regime increasingly does not distinguish between moderate and radical forms of Islamic activism. The fact that moderate Islamic activists are integral to civil society has furthermore meant that, as a consequence of the shift in government policy, civil society as a whole has been under assault from the government.

In 1993 the government declared the creation of a new professional syndicate law, called the Law for Democratic Guarantees in Professional Syndicates (Law 100 for 1993). Based upon the rationale that the Brothers are winning the elections only because an active minority of supporters are participating in the elections, the law requires a minimum quorum of voters for the results to be held as legal. When the Islamists recently proved that they were capable of winning syndicate elections under the new law, the government invoked various pretexts to postpone elections in syndicates dominated by Islamists.[26]

Similarly, in 1994, after the success of the Islamists in various local and associational elections, the government also enacted legislation to terminate village mayoral elections and university elections. Academic deans in Egypt's universities and mayors are now appointed, ending 20 years of popular voting.[27]

Changes to the Law of Political Parties mean that the government can also now more easily disqualify people from running as candidates for posts in parties or in national elections.[28] In a further effort to prevent the success of the Islamic option in the 1995 elections, hundreds of Brothers were arrested in the two days prior to the elections.

Under the force of public protest and pressure President Mubarak was forced to withdraw Law 93 of 1995, a new press law that stiffened penalties for publishing materials that (loosely defined) were in contempt of or decried government official or state institutions.[29] However, the government continues to censor and restrict the press.

These changes have occurred in conjunction with amendments to the penal code that strengthen the repressive capacity of the state. Beginning in 1992 the government established a series of military courts that are more efficient than the regular bogged-down court system and offer no right of appeal. Human rights activists complain that, as part of official policy, thousands of political prisoners are being systematically tortured and hundreds of Islamic activists have disappeared.[30]

Egypt's NGOs have not remained unaffected by either Egypt's structural reform program or the struggle between radical Islamists and the state. As with other forms of civil society, Egypt's NGOs have been restricted by increasing state control over the past five years. This control has hindered their ability both to deal adequately with those negatively affected by structural reform and to increase their political influence in the realm of social welfare as the state economically withdraws.

NGOs in Egypt: service capacity and sustainability

All NGOs in Egypt must register with the Ministry of Social Affairs (MOSA). Of the NGOs dealing in the field of social welfare, this chapter will treat them according to three types: religious (primarily Islamic) and secular social-welfare NGOs, and Community Development Associations (CDAs). CDAs differ from other NGOs in that they are working in the broader field of community development, providing a village or urban neighborhood with the basic social services. They therefore carry out more activities than other NGOs. Many CDAs were established by the government (or taken over) during the Nasser era, and were staffed and funded directly by the government with a view, at least from the government's point of view, to implementing the state's development program.

Religious social-welfare NGOs are those that are affiliated with a mosque or church. They are located beside or in the basement of the mosque and may offer a whole range of social-welfare services, the most popular of which are health-care and day-care. These NGOs are not run by the mosque or church *per se*, but by law have their own independent boards of directors.

The number of civil associations (all types of CDAs and NGOs) registered with the MOSA is 14 500.[31] This figure represents an impressive three-fold increase over the past 30 years.[32] Amongst other services these NGOs provide financial assistance and health services to approximately 5 million and 14 million Egyptians respectively (not accounting for multiple-usage).[33] NGOs have become so fundamental for Egyptian society that in, certain cities and villages, entire quarters depend on their services.[34] Government services are cheap and often even free, but they are overcrowded, inefficient and often of poor quality. Despite recent efforts to rectify the situation, the Mubarak government cannot provide the social-welfare services needed by its rapidly growing population. NGOs offer good, inexpensive and accessible social-welfare services that the government seems unable to provide.

Despite their noteworthy record, however, Egyptian NGOs suffer from a number of weaknesses. In fact a large percentage of NGOs are either defunct or doing very little. NGO representatives and international donors are generally in agreement concerning the weaknesses NGOs must overcome in order to better address the increasing needs of the poor and middle classes. The most common complaints concern the fact that many NGOs are overstretched both in their ambitions and their trained staff. This means that plans are often not put into practice and that NGO service delivery can be inconsistent and unreliable.

Common problems also include poor organizational skills and financial reporting. International donors note that some of even the largest and best NGOs suffer from weak organizational structure with one dynamic person running the entire operation. This often results in a paternalistic and/or authoritarian mode of decision-making. In the end both the sustainability of the NGO and its ability to accurately reflect the needs of its surrounding community comes into question.

Finally, Egyptian NGOs do not form what can be labeled an "NGO movement." There is little or no coordination between NGOs or what can be called NGO coordinating bodies. NGOs are therefore not able to address or target societal needs in any coordinated or planned fashion. This weakness is due both to the poor organizational structure of many NGOs and the present political climate of confrontations, politically and militarily, between the state and militant Islamic forces. With the exception of a few powerful and/or protected groups, most attempts at creating independent NGO coordinating bodies have been thwarted (or the associations have been dissolved) under the Law of Association of the Emergency Laws which have remained in effect since President Sadat's assassination in 1981.[35]

NGO funding and state dependence

According to a 1994 World Bank Study on NGOs in three governorates, estimated total revenues of Egypt's NGO sector in 1992 were US$85–110 million and its financial reserves were US$22–30 million.[36] These figures do not represent a sector which is unable to mobilize resources and revenues. Contrary to popular belief the Egyptian NGO sector, as a whole, does not suffer from a lack of financial resources. The study furthermore reveals that the bulk of this financing does not come from state or international aid; rather it is primarily a result of private donations and activity (user) fees.

Despite the optimism of the World Bank study, if we break down these figures there are important imbalances in the distribution of these funds and revenues. Other than the quality, and hence desirability, of services, these imbalances are due to the geographic location, type, and size of the NGO.

NGOs in urban areas are able to attract a larger clientele and are able to ensure themselves a greater percentage of activity fees than rural NGOs. They simply have a larger "market."[37] In addition, rural NGOs have difficulty in attracting and keeping qualified staff. Doctors, for example, are disinclined to work in rural areas where relatively few people can afford their services as compared to the urban areas. This, of course, detrimentally affects the number of users. For similar reasons, urban NGOs generally collect far greater sums in donations. This is supported by the World Bank study which found, for example, that, while a small number of metropolitan and urban NGOs are able to collect major sums, a large percentage of NGOs collect no donations at all.[38]

If we examine NGOs according to type, we find that while (secular) social welfare NGOs, religious NGOs and CDAs all primarily rely upon their own resources – activity fees, private donations and other sources – for most of their income,[39] there are some important differences. Religious NGOs are far more independent of state aid and other grants than secular social-welfare NGOs and, in particular, CDAs.

Activity fees make up approximately 52 percent, donations 34 percent, and state and foreign aid 9 percent of total revenues of religious NGOs. In terms of secular social welfare NGOs, donations represent approximately 24 percent of their revenue, while activity fees, and state and foreign aid represent approximately 22 percent and 20 percent, respectively. Finally, the percentage of activity fees, state and foreign aid, and donations of total revenue of CDAs is approximately 31 percent, 34 percent and 15 percent, respectively.[40]

Financially, religious NGOs have the advantage of the mosque from which *zakat* (alms) flows to the associated NGO. Donations designated for the mosque are also often redirected to the NGO.[41] CDAs and social-welfare NGOs do not enjoy these advantages. This contributes to religious NGOs' ability to be more self-reliant.

According to its figures, MOSA provides aid in the form of funds, personnel, and technical and training services, to approximately 35 percent of Egyptian NGOs. While many CDAs do not receive any state assistance, this aid remains generally concentrated on a relatively small number of CDAs as opposed to other NGOs. In these cases the CDAs are dependent on the state for over 30 percent of the total revenues.[42] Whether because the majority of CDAs are in the rural areas or because MOSA favors provincial towns and urban centers, the bulk of these CDAs are also located outside of the capital.

Foreign aid, dominated by USAID, is also generally concentrated in the governorates of Upper Egypt, where poverty is deemed to be most prevalent, rather than in the capital. However, given the increase of poverty in, and the increased migration to Cairo, donors are noting the need to reexamine this strategy.

Foreign aid, furthermore, is generally concentrated on a relatively small number of very large, effective and generally older NGOs. A small handful of NGOs, such as CEOSS, the Coptic Evangelical Organization for Social Services, are the favored beneficiaries of foreign aid for a variety of reasons. The most important of these reasons is the simple fact that donors are dealing with large sums of money which they would like to ensure are effectively spent and implemented. As the above discussion of NGO organizational weaknesses indicates, the majority of NGOs simply do not reach international donors' standards.

In fact, many donors complain of what they see as a gap between a small number of NGOs which are flooded, sometimes to their detriment, with donor aid, and the majority of NGOs which often struggle to exist. The new Social Fund for Development established by the World Bank in order to alleviate some of the negative effects of structural adjustment at the microlevel, for example, has been accused by international donors of dispensing large sums of money with little discretion or monitoring. Many NGOs do not have the experience, capacity or management to deal with large sums of money and implement the desired programs. They are limited by their personnel and overextended.

The result is a sector which is dominated by weak, low-income NGOs. Only 65 percent of registered NGOs are active and community-oriented (as opposed to private membership associations) and of these, 25 percent are very weak and, in many cases, active only part-time and 27 percent have limited success as providers of one or two small but regular services and one or two minor, periodic activities.[43]

At the other end of the scale, only 5 percent have one to four major service programs, expansive facilities and institutionalized relationships to the local community. These are strong and active organizations with full-time programs and administrative support staff.[44] Without meaning to discount the financially secure medium-range NGOs (and particularly the religious NGOs), the majority of successful NGOs are either highly dependent upon state aid and primarily located in rural areas, or situated in the more affluent areas of Cairo where users have the means to pay and/or donate for services. In this scenario the increasing numbers of urban poor are the most hurt. Smaller NGOs in poorer areas are caught in a vicious cycle of limited or no state or foreign aid, few or no influential board members with helpful connections in the government, a lack of trained personnel, an increasing number of patrons without the means to pay, and ultimately limited effectiveness.

The state's political control over NGOs: MOSA and Law 32

While the state has withdrawn economically from the socioeconomic sphere, this has not translated into political withdrawal. The increased financial and programmatic burdens on NGOs are not occurring within the context of increased political freedoms to improve their financial or decision-making capabilities or to influence policy. The state is simply not creating the necessary political space.

As stated above, all NGOs in Egypt must register with MOSA. MOSA's official function is to encourage individual efforts in the social domain. Established in 1939, its goals include collaborating with associations, as well as encouraging, orienting, financially aiding and coordinating them, and raising their standard of social work. However, in reality, MOSA's role is more to regulate and monitor the associations.

This regulatory approach towards NGOs is evident in Law 32, issued under Nasser in 1964, by which all NGOs today must still abide.[45] Law 32 legalizes the government's, specifically MOSA's, control of NGOs in Egypt. Under Law 32, MOSA is given wide powers of discretion to rule on the right of an association to exist. MOSA is able to dissolve an association without asking a court to carry out its orders. While associations may appeal, it is only to an administrative authority within

MOSA and they can do so only if their application is rejected, not if they are dissolved. MOSA can, furthermore, oppose any candidate for the administration council of an association. It can nominate representatives from MOSA, up to 50 percent, for an unlimited time period to the council, and it can dissolve and nominate new councils.

Registering an association as stipulated by MOSA is quite onerous and entails at least ten bureaucratic steps, public notarizing of all founders' signatures, a written by-law, and the approval of at least three ministerial bodies including the Ministry of Interior's General Security Department. The entire procedure is frustrating, time-consuming and without any guarantees.

Once registered, NGOs labor under a variety of other restrictions. Ibrahim's 1994 study of the impact of Law 32 on Egypt's NGOs and private foundations finds five categories of constraints which MOSA imposes upon NGOs: administrative, geographic, substantive (activities), financial, organizational, and technical.[46] Law 32, for example, imposes stringent requirements in terms of its record-keeping, recording and inspection of NGOs. These constraints furthermore include limitations on membership, fund-raising, beneficiaries, and the scope of activities.

In terms of fund-raising, for example, Law 32 prohibits receiving money from a foreigner and/or a foreign organization without the prior permission of MOSA. It also does not permit NGOs to collect donations from the public; organize shows, bazaars, sporting events, or any other fund-raising activities except by permission of MOSA. A maximum of two requests for fund-raising campaigns may be granted. Law 32 furthermore places restrictions on investing money.

Violating the provisions of Law 32 can result in penalties, as well as a prison term for up to six months. Activist non-governmental organizations which have incurred the displeasure of the authorities have also been prosecuted under a wide range of legislation which limits freedom of expression in Egypt, and have been detained under emergency legislation. Hence, the usefulness of Law 32 for inhibiting the development of NGOs cannot be underestimated.

Another means of MOSA control is via the designation of "public-interest" status upon certain NGOs. NGOs which reach a certain size and level of effectiveness may be placed in this category. As such they receive certain privileges and funding. A high concentration of MOSA's financial aid goes to these public-interest NGOs.[47] It also is, however, an effective means of control as these NGOs generally become a mere extension of the state, implementing MOSA's own projects.

Many NGOs do operate in an uninhibited fashion from MOSA and Law 32. Their abilities to raise funds and provide services are generally unimpeded. However, these NGOs either have connections in MOSA, powerful people on their boards of directors, or MOSA employees on their boards of administration. Indeed, of the above-mentioned 5 percent of NGOs which represent Egypt's largest NGOs:

> one-half ... are social service associations with close historical ties to government (usually the Ministry of Social Affairs) and directed by former (retired) senior ministry officials with government-seconded staff.[48]

While many MOSA officials who have been placed on NGO boards have proven to interfere and place obstacles in associations' ways, others have proven to be highly useful to NGOs, helping them through government regulations and bureaucracy. Some officials have simply proven to be quite financially costly for NGOs, demanding various forms of income and payment for attending meetings.

A good example is provided by one of the larger and older Islamic NGOs in Sayyida Zeinab, Cairo, in which the director is also a retired general within the police with numerous connections within the government. In addition, several ministers are members of the association. In an interview with the author the director stated: "The Association of '*x*' is not to be toyed with. I'm a general in the police, graduated in 1954, and the Minister of the Interior was in the same graduating class as me and we also have support of four governors: Port Said, Mansura, Asyut and El Minya. The governor of El Minya was a bunk-bed partner of mine." In other words, a large degree of an association's success depends on the members' ability to get around impediments through informal arrangements and networks which incorporate state officials.[49]

Despite the exceptions, those NGOs that manage to operate unhindered by MOSA and Law 32, the restrictions and regulations imposed by MOSA and Law 32 are problematic enough to force some NGOs to operate illegally or to register under a different legal form. A study carried out by the Ibn Khaldoun Center for Development Studies in 1993 found that there were on average seven informal grassroots initiatives for every formally registered NGO. In addition, many NGOs now opt for registering as civil or professional companies instead of registering as NGOs and falling under the control of Law 32.[50] As companies they are liable to taxation on their revenues. However, they feel this is preferable to the arbitrariness of Law 32 and of MOSA bureaucrats.

International donors have also recognized the need to circumvent MOSA. Many donors are willing to fund civil companies, and Oxfam, for example, encourages NGOs of all types to register as companies. Other donors, the Netherlands Embassy for example, refuse to inform MOSA of all of their grants (much less funnel their monies via MOSA as is stated by law) and choose to give their monies directly to NGOs. It is then the NGOs' responsibility to acknowledge their receipt of money or not to MOSA, according to Law 32. These donors refuse to participate in what they see as the policing arm of MOSA. They furthermore have greater assurance that the funds will arrive at the desired destination.

Yet MOSA is not the only ministry to exert control over NGOs. Depending on the services they offer, NGOs may have to register and comply, for example, with the regulations of the ministries of Education, Health or Culture. In an attempt to avoid these regulations, the above-mentioned World Bank study notes, for example, that many religious schools, primarily Islamic, are established as an "effective NGO strategy for circumventing bureaucratic obstacles and garnering private donations for multi-purpose facilities."[51] In other words they are able to avoid the burdensome and expensive regulations and building requirements of the Ministry of Education, and take advantage of the Ministry of *Awqaf* and the University of *al-Azhar* grants for religious education institutions and mosques. Filing "mosque construction" on the permits for soliciting donations also facilitates approval.

These forms of state control are not lessening with economic reform and may in fact be increasing. An excellent example of this is the fact that there are three bodies dealing with NGOs in the field of population and family planning. One is the original official governmental body, the Egyptian Society for Population and Development. The second is an advisory committee appointed by the Ministry of Health and Population. The third, the National NGO Committee for Population Development (NCPD), is a coalition of NGOs struggling to avoid ministerial control both by the Ministry of Health and Population and by MOSA. As this latter group, for example, comprises NGOs dealing in population and family planning, the two ministries are at odds as to under whose jurisdiction they fall.

The existence of three bodies (and to a large extent, their composition) is largely a result of what can only be called ministerial turf wars. NGO representatives and donors fear that as ministries are increasingly affected by privatization and, as a consequence, their areas of jurisdiction are either eliminated or transferred elsewhere, the efforts to gain or maintain control over the NGO sector by various ministries will intensify.

Furthermore, the government, and MOSA in particular, uses the present political climate of violence between militant Islamic groups and security forces as a justification for the continuation of Law 32. Since the onset of Islamic militancy, it has in fact been enforcing the law with increased vigor. MOSA argues that deregulating Law 32 or loosening its grip over NGOs would only serve the interests of Islamic extremists who would then be able to establish NGOs as covers for illegal activities against the government and the state. The government has thus made increasing use of Law 32 to intimidate and suspend various associations.

In the context of economic reform, therefore, political space has been decreasing for Egypt's NGOs. Yet ironically, many NGOs complain that Islamic NGOs are the least affected in their capacity to provide for the poor in the context of the recent crackdowns. Article 17 of the By-Laws of Presidential Decree 932 of 1966, for example, gives religious NGOs the right to collect donations in places of worship without prior permission from MOSA, a privilege denied other NGOs.[52] This exemption from the restrictive fund-raising elements of Law 32 may perhaps be explained by the fact that the government warily supports Islamic services because they satisfy the poor and middle classes, and they limit the numbers seeking help from the overburdened state.[53] Although suspicious of Islamic associations, it sees in them, at least until recently, an opportunity both to gain credit for their successes and discredit Islamic groups' claims that the state is antagonistic toward Islam. Opponents to Law 32 therefore argue that Law 32 cripples their success both in absolute and relative terms compared to Islamic NGOs. Certainly, this is attested by the fact that Islamic NGOs are the least dependent upon state and foreign aid in comparison to other NGOs and CDAs. Their unique ability to raise funds plays a large role in this success.

Law 32 and Egypt's democracy: Law 32 under review

The political impact of Law 32 is felt far beyond those working in NGOs. At best, the rigorous enforcement of Law 32 erodes the government's positive identification with groups which have gained local support and respect. At its worst the government is seen as hostile not only to Islamic groups and associations (moderate and militant) but to the population at large.

The 1992 earthquake provides an excellent example. As volunteer associations, and particularly Islamic associations, rushed to help earthquake victims, the government prohibited activities such as fund-raising and other forms of assistance except through the government or the Egyptian Red Crescent. The main reason behind the government's decision was to weaken the activities of Islamic groups, and to deny them the opportunity to criticize the government through their aid activities.[54] The government, however, was unable to provide the necessary services.

By preventing people from helping themselves, this sends the message that the government – and potentially the (so-called) democratic state – is the enemy. Opposition leaders are united in their concern that the assaults on civil society via Law 32 and other measures are undermining the democratic option in Egypt and forcing Egyptians, particularly moderate Islamists, to seek a militant option.

For this reason, and the basic fact that it denies democratic freedoms to Egyptians, Law 32 has become a source of contention in the struggle for greater democratization in Egypt. It stands in contradiction to the 1971 Constitution which guarantees the right of association, as well as the International Covenant on Civil and Political Rights (1982) which Egypt ratified.[55] Hence representatives from the entire political spectrum have spoken out against the law as detrimental to political pluralism and democracy in Egypt.[56] It has been raised in parliament and is regularly criticized in the opposition press.

These criticisms, as well as the continuation of Islamic violence, have led the government to review the effectiveness of Law 32 and the premise that the control of NGOs is necessary to curb unrest. In late 1997, under a new Minister of Social Affairs, Mervay al-Talawy, the government set up a committee to draft a new law regulating NGO activities. The committee was composed of approximately 30 legal experts, civil society activists, and intellectuals. As of yet, a new draft law has not been presented to parliament. NGOs are involved in on-going efforts to ensure that any new law reflects their concerns and interests.

Conclusions

Social-welfare NGOs in Egypt provide an important "cushion" between state and society which absorbs some of the impact of painful economic restructuring on the poor and middle classes. The need for this cushion is becoming greater as the government more seriously embarks upon its privatization program. As has been shown, however, this does not mean that NGOs will necessarily be financially, organizationally or legally able to continue performing this cushioning role or to fill the "gap" the state leaves behind.

Politically, unless the government changes Law 32, it will continue to control the establishment, expansion, functioning and administration of NGOs, thereby limiting their ability to create a greater policy-making role for themselves. Since 1990, opposition members within parliament and various advocacy NGOs have brought increasing pressure upon the government to amend Law 32. This has resulted in the drafting of a potential new law.

However, while the abolition of Law 32 would increase NGOs' ability to raise funds and their political autonomy, it would not be enough. Law 32 is not the only means through which civil society is controlled in Egypt. Furthermore, ministers and bureaucrats are not willing to relinquish power and its associated perks. In addition, despite economic reforms, Egypt's corporatist structure remains fully intact. Avenues to greater political participation and policy-advising have not been opened.

Moreover, economic reform is taking place in Egypt today amidst little or no discussion concerning the future role of the state or of NGOs in the field of social welfare. On their own, NGOs cannot provide the over-arching body, planning or management in an important, large and complex field such as social welfare, which includes health, day-care and education. Were the state to withdraw, NGOs cannot be expected to simply "fill the gap" and take over the state's role. A discussion of some form of division of roles and powers is sorely needed in Egypt's economic reform.

The case of Egypt's social-welfare NGOs points to the fact that a more exclusionary authoritarianism has accompanied economic reform in the country. The increasing civil and human rights abuses in the security forces' campaign against Islamic militants is one aspect in this tendency. Included in this list of exclusionary tactics can be added the government's actions against Islamic aid groups after the 1992 earthquake and, in particular, the newly created Law 100 which regulates elections within the syndicates (to the government's benefit). Certainly, Egypt indicates that political opening will come about only through planned political reform, not just economic reform.

Notes

1 For the purposes of this chapter, corporatism is defined as "a system of interest representation in which the constituent units are organized into a limited number of singular, compulsory, non-competitive, hierarchically ordered and functionally differentiated categories, recognized or licenced (if not created) by the state and granted a deliberate representational monopoly within their respective categories in exchange for observing certain controls on their selection of leaders and articulation of demands and supports." Robert Bianchi, *Unruly Corporatism: Associational Life in Twentieth-Century Egypt* (New York: Oxford University Press, 1989), p. 226, footnote. 14.

2 Ibid., p. 86.

3 See Ann M. Lesch, "Democracy in Doses: Mubarak Launches his Second Term as President," *Arab Studies Quarterly*, **11**, 4 (1989), 90; Nadia Hijab, "Reform in the Wake of Egypt's Elections," *The Middle East*, **117** (July 1984), 23.

4 In June 1986 Egypt's foreign debt was estimated at about $40 billion. Said el-Naggar, "Prospects and Problems of Privatization," in Ibrahim M. Oweiss (ed.), *The Political Economy of Contemporary Egypt* (Washington, DC: Center of Contemporary Arab Studies, Georgetown University, 1990), p. 62.

5 Hans Lofgren, "Egypt's Program for Stabilization and Structural Adjustment: an Assessment," *Cairo Papers* **16**, 3 (1993), 22.

6 Ibid.

7 Daniel Brumberg, "Survival Strategies versus Democratic Bargains: the Politics of Economic Reform in Contemporary Egypt," in Henri Barkey (ed.), *The Politics of Economic Reform in the Middle East* (New York: St Martin's Press, 1992), p. 90.

8 El-Naggar, "Prospects and Problems of Privatization," p. 50.

9 John Waterbury, "The Political Context of Public Sector Reform and Privatization in Egypt, India, Mexico, and Turkey," in Ezra N. Suleiman and John Waterbury (eds), *The Political Economy of Public Sector Reform and Privatization* (Boulder, Co.: Westview Press, 1990), p. 296.

10 El-Naggar, "Prospects and Problems of Privatization," p. 63. See also John Waterbury, *Exposed to Innumerable Delusions: Public Enterprise and State Power in Egypt, India, Mexico, and Turkey* (Boston, Mass: Cambridge University Press, 1993), p. 141.

11 El-Naggar, "Prospects and Problems of Privatization," p. 63.

12 Ibid., p. 50. El-Naggar notes how other factors also contributed to the poor performance of the Egyptian economy but that the public sector remains of major importance.

13 Denis Sullivan, "Extra-State Actors and Privatization in Egypt," in Iliya Harik and Denis J. Sullivan (eds), *Privatization and Liberalization in the Middle East* (Indianapolis: Indiana University Press, 1992) p. 26

14 Saad Eddin Ibrahim, "Governance and Structural Adjustment: the Egyptian Case." Unpublished paper, Ibn Khaldoun Center for Development Studies, Cairo, November 1994, pp. 15–17, and his "The Year in Review," *Civil Society*, **7**, 74 (February 1998), 14–17.

15 Cassandra, "The Impending Crisis in Egypt," *Middle East Journal*, **49**, 1 (Winter 1995), 11–13; Amos Elon, "One Foot on the Moon," *New York Review*, 6 April 1995, 34; Saahir Lone, "Government Cracks Down on Militants," *Civil Society*, **4**, 39 (March 1995), 9; Simon Brindle, "Egypt: a High Rise Economy," *The Middle East* (January 1995), 25–26; Caryle Murphy, "The Business of Political Change," *Current History*, **94**, 588 (January 1995) 19.
16 "The Year in Review," *Civil Society*, **7**, 74 (February 1998), 11.
17 Egyptian Preparatory Committee for the 1995 International Women's Forum, *Report of the Egyptian NGOs for the Forum on Women, Beijing 1995: Egyptian Women's Status from Nairobi to Beijing* (Cairo: The Alliance for Arab Women, 1995), p. 28.
18 The World Bank, *World Development Indicators 1997* (Washington, DC: The World Bank, 1997), p. 78.
19 The World Bank, *Social Indicators of Development* (Baltimore, MD: Johns Hopkins University Press, 1996), p. 103.
20 Ibid., p. 102.
21 Arab Republic of Egypt, *Statistical Yearbook* (Cairo: General Agency for Public Mobilization and Statistics, 1994), p. 166.
22 Robert J. LaTowsky, "Financial Profile of Egypt's PVO Sector." Unpublished paper, World Bank, Washington, DC, June 1994, p. 8; Amani Qandil and Sarah Ben Nefissa, *Civil Associations in Egypt* (Cairo: Al-Ahram Centre for Political and Strategic Studies, 1994), p. 111.
23 Saad Eddin Ibrahim, "The Changing Face of Islamic Activism," *Civil Society*, **4**, 41 (May 1995), 7; S. Lone, "Economic and Social Roots of Religious Movements," *Civil Society*, **4**, 41 (May 1995), 23–4; Cassandra, "Impending Crisis," pp. 9–27.
24 Ibrahim, "Changing Face," p. 8.
25 Ibid., p. 6; Cassandra, "Impending Crisis," p. 20.
26 Mustapha Kamil al-Sayyid, "A Civil Society in Egypt?," in Augustus Richard Norton (ed.), *Civil Society in the Middle East*, vol. I (Leiden: E. J. Brill, 1995), pp. 285–6; see also "Egypt: Parliament Pushes Union Law Through," *Civil Society*, **15** (March 1993), 3–6 (reprinted from *Al-Ahram Weekly*).
27 Cassandra, "Impending Crisis," p. 16; Ibrahim, "Civil Society and Electoral Politics in Egypt," *Civil Society*, **4**, **48** (December 1995), 6.
28 al-Sayyid, "A Civil Society in Egypt?," pp. 283–4.
29 Ibrahim, "Egypt's Civil Society Challenged," *Civil Society*, **4**, 43 (July 1995), 4; "Press Law Invokes Anger," *Civil Society*, **4**, 42 (July 1995), 6; "The End of Law 32," *Middle East Times*, 23–9 June 1996, p. 1.
30 Mary Ann Weaver, "The Novelist and the Sheikh," *The New Yorker*, 30 January 1995, p. 57; "Sad Dreams by the Nile," *The Economist*, 3 June 1995, p. 35.
31 LaTowsky, "Egypt's PVO Sector," p. 32. Estimates of civil associations vary greatly due the difficulty of data collection and inaccurate government documents. In addition, surveys use different categorization systems. Latowsky categorizes civil associations into six types. Qandil and Ben Nefissa use a typology of 17 types and estimate the total of civil associations (including all types of NGOs and CDAs) as 11 313. See Qandil and Ben Nefissa, *Civil Associations in Egypt*, p. 111.

32 Saad Eddin Ibrahim, "Law 32 of 1964 on Egypt's PVOs and PFs: a Critical Assessment." Unpublished paper, the Ibn Khaldoun Center for Development Studies, Cairo, September 1994, p. 4 of the Executive Summary.
33 Ibid., p. 51.
34 Sarah Ben Nefissa-Paris, "Le mouvement associatif egyptien et l'islam," *Maghreb-Machrek,* 135 (January-March 1992), 34.
35 Cassandra, "Impending Crisis," p. 15. President Mubarak renewed the Laws for an additional three years in 1994. Exceptions include powerful businessmen's associations and the National Council for Population Development which was formed in preparation for an international (and highly publicized) conference on population held in Cairo.
36 LaTowsky, "Egypt's PVO Sector," p. 2.
37 Ibid., p. 11.
38 Ibid., p. 14.
39 Ibid., p. 27.
40 Ibid., p. 26.
41 Ibid., p. 14.
42 Ibid., p. 19.
43 Ibid., p. 32.
44 Ibid., p. 33.
45 This section is based on excerpts from a report published by the Middle East and North Africa Program of the Lawyers Committee for Human Rights and the Egyptian Organization for Human Rights, December 1991, "Restricting the Human Rights Movement in Egypt: Legal Restrictions on Independent Non-Governmental Organisations," *Civil Society,* 5 (May 1992), 6. See also Sarah Ben Nefissa-Paris, "L'etat egyptien et le monde associatif," *Egypte/Monde Arabe,* 8 (1991), 118.
46 Ibrahim, "Law 32," pp. 23–5.
47 Ibid., pp. 48–9.
48 LaTowsky, "Egypt's PVO Sector," 33.
49 Sami Zubaida, "Islam, the State and Democracy," *Middle East Report,* 179 (November–December 1992), 7.
50 Ibrahim, "Law 32," pp. 53, 62.
51 LaTowsky, "Egypt's PVO Sector," p. 17.
52 Ibrahim, "Law 32," pp. 57–9.
53 *Al-Wafd,* 12 April 1987.
54 "A One Minute Earthquake," *Civil Society,* 11 (November 1992), 3.
55 "Restricting Human Rights Movement," p. 7.
56 Ben Nefissa-Paris, "Le mouvement associatif," 31–2; idem., "L'etat egyptien," p. 119.

8

Business–State Relations after Liberalization in Jordan[1]

Pete W. Moore

Introduction

In the mid-1980s Jordan began steps toward economic restructuring and reform. In 1989 the state formally introduced political liberalization by holding the first parliamentary elections in over 20 years. These events, and analogous ones in the developing world, generate expectations for changes in the way classes and groups deal with the state. For this chapter, the focus is on change in state–business relations. There are generally three expectations regarding this relationship and its potential impact on political liberalization/democratization. First, state disengagement from its dominant economic role allows greater policy involvement by the private sector and its associations. Private sector elites then utilize this access to block further political liberalization that would endanger (i.e. greater access for labor) their capital positions.[2] A second expectation suggests – whatever business's political intentions – that the greater accountability won through increased private sector involvement can have "positive externalities" for other social groups.[3] Accountability is, in other words, fungible. Third, state divestment leads to a closer and more amiable relationship between the state and the private sector. This new pact would lead to increased private sector influence over national economic policy, providing crucial support for the economic future of political liberalization.[4]

The question then is what variables figure into shaping these outcomes. This chapter addresses this question by discussing the history of business–state relations in Jordan and how this relationship fared after Jordan took its first steps toward political liberalization in the late 1980s. The argument is that none of these outcomes (democratization blocked, unintentional accountability, or business–state pact) has transpired in the Jordanian case. The role of organized business (that is business associations) has been a key factor. Weak institutions of business representation have been unable to advance consistent and productive policy input that would allow business a greater voice in policy or facilitate a new pact between the private sector and the state. In other words, despite over a decade of efforts at economic and political liberalization in Jordan, business has not been able to take advantage.

For Jordan the stagnation of the economy in the late 1980s demanded greater coordination between the private sector and the state as one possible solution. Consequently, greater political openings after 1989 might have allowed business a more significant role in economic policy-making. Patterns of relations after the onset of political liberalization in 1989, however, reveal a mixed outcome. Business representatives have certainly been present to a greater degree in national debates over economic policy, but coordination remains shallow and a new pact between the state and the private sector for Jordan's economic future has yet to emerge. Indeed, fundamental obstacles impeding effective business lobbying have been strengthened rather than alleviated in the post-liberalization environment. The fate of business–state relations in Jordan takes on added significance given its role in the Middle East peace process. Moreover, Jordan is a good test case in a region where states facing similar economic troubles have turned to economic and political liberalization as a means to cope. Unlike more advanced political liberalizations in Latin America and Eastern Europe, many Arab states (e.g. Egypt, Kuwait, Morocco, and Jordan) appear stuck in "half-way houses"[5] of reform, neither moving forward to more expansive political liberalization nor solving their economic woes.

The issue of business–state relations in the case of Jordan also offers a useful window on post-rentier politics in the developing world. The central premise of rentier theory is that the character of a state's revenue is basic to a country's politics.[6] Therefore, whether the bulk of a state's revenue is derived internally (through forms of taxation) or externally (through rents) has fundamental implications for how state and society will interact. Like many states in the developing world in the 1970s, the Jordanian state feasted on exogenous revenue in the form of foreign aid (for other states this revenue came in the form of oil and mineral concessions). Where these finances accounted for over half of state revenue, as was the case in Jordan and many Arab states, the effects were quite uniform. Fiscally buoyant states drove their domestic economies through distribution of these rents and coopted important social groups, such as business, through widespread patronage. During the high oil prices of the 1970s this strategy served short-term economic and political goals. Once these rents began to decline, in the 1980s, the basic expectation was that formally coopted groups should be able to reassert policy influence with fiscally vulnerable state officials. Actual patterns of state–society relations in the 1980s and 1990s across the Middle East have not conformed to the basic expectation. Various degrees of political and economic liberalization have taken place but previously coopted groups have failed to pressure states into genuine power-sharing. Useful revisions to the basic rentier theory have contributed to our understanding of why post-rentier outcomes have varied.[7] The empirical discussion in this chapter contributes to the on-going revision of rentier theory by examining the fate of business–state relations in a post-rentier environment. In this way it is clear that post-rentier politics and the future of political liberalization in countries such Jordan are tightly wound.

The next section proceeds by briefly introducing the history of business–state relations in Jordan. The third section discusses the political and economic changes that accompanied political liberalization in 1989 and then reviews three issues in the 1990s involving business–state relations. The fourth section presents the conclusions.

Business–state relations in Jordan

As a result of agreements between France and Britain, what was then known as TransJordan came under British mandate in 1920. Thwarted from a seat in Damascus, Emir Abdullah Ibn Hussein agreed to British support of a Hashemite throne in Amman. Independence from Britain came in 1948. In the period prior to and immediately after independence, business elites and their organizations enjoyed moderately influential relations with King Abdullah. As rents flowed to the state in 1970s, this relationship changed and business influence in economic policy declined. Throughout the 1980s a decrease in rents combined with ongoing structural change forced a reassessment of state strategies toward the business community. These shifts set the scene for economic and political liberalization in the 1990s.

The region of TransJordan that Abdullah came to rule was predominantly rural and tribal in character. Trade and commerce were not highly developed. As British–Hashemite political authority took root in Amman, urban merchants, who had originally moved from Nablus in Palestine and Damascus in Syria to the cities of Irbid and Salt, gradually took up residence in Amman. Thus, in many ways Abdullah and the Syrian and Palestinian merchants that followed were all strangers to a new land. Together they built an economy from scratch and from this shared endeavor a close relationship resulted. Merchants and rulers in the pre-independence period in Jordan interacted in ways very similar to other parts of the Middle East. Much like Jill Crystal's account of merchant families and the ruling al-Sabahs in Kuwait, the Hashemites turned occasionally to leading merchants for *ad-hoc* funds for state activities.[8] These were also very active merchants, establishing the Amman Chamber of Commerce (ACC) in 1923 as a means to organize and regulate dealings with the state. Of the nine elected presidents of the Chamber, all have come from merchant families from either Nablus or Damascus, that is, merchant families who came to Jordan before 1948.[9] These leaders and their community enjoyed two advantages.

First, many of the original merchant families established their business prominence by securing monopolistic trading rights in Mandate Palestine from British officials. Some of the leading merchants received support from the monarchy in winning these concessions from the British. In return, leading merchants provided funds to the monarchy, which otherwise depended exclusively on transfer payments from the British.[10]

Structural homogeneity was a second strength which the merchant community enjoyed in these years. These were traders, importers, and re-exporters to local and regional markets. The interests of both large and small traders more or less converged on many issues. Such advantages aided organized business, such as the ACC, in lobbying the state in the 1950s and early 1960s. For instance, government proposals to institute an income tax in the 1950s were turned back, due in part to successful lobbying by the ACC. And in the early 1960s the Chamber led domestic opinion in Jordan convincing the government to participate in the ill-fated Arab Common Market.[11] By this period, though, changes were under way that would bring an end to this close association.

Domestic and regional trends that impacted many Arab countries in the 1960s also altered established domestic political arrangements in Jordan. The 1967 War was a watershed for social and economic change. The influx of Palestinian refugees from the War provided a wider base for a new entrepreneurial class. Though still a small country, Jordan became more diverse. Membership of the ACC significantly increased in the years after 1967 as Palestinian refugees opened businesses.[12] Economic changes accompanying the War reinforced these tensions.

Two aspects of the new economic environment were key for state–business relations: the dramatic increase of external finance to the government budget, and remittances from workers in the Gulf. From 1966 to 1967 direct external aid to the Jordanian government jumped from just 9.9 million Jordanian dinar (JD) to over JD40 million. That represented over 50 percent of government revenue for that year.[13] Increased state resources expanded strategies for coopting the business community. One avenue pursued was state investment in or establishment of semi-public companies. Beginning in the 1960s, government creation and investment in previously private companies gave state elites a powerful tool for rewarding or punishing business elites. Various ministries and later the Jordan Investment Corporation (JIC) steadily bought up shares in leading companies. With majority positions on the executive boards of companies such as the Arab Potash Company, Jordan Cement Companies, Jordan Phosphates Mines, and Royal Jordanian Airlines, the state could appoint most of the positions of the board.[14] These appointments along with lucrative government contracts and cheap real-estate sales gave state elites new, powerful levers to coopt some business elites.

A second state strategy was the effort to construct a wholly new business class, one dependent upon state patronage. The target was to induce sectors of East Bank elites and pre-1948 Palestinians to launch industrial and business investments. Thus, a rival to the expansive business community of Palestinian traders and importers centered in Amman could develop. In 1965 the Industrial Development Bank was created to extend start-up loans for the nascent industrial sector. A series of laws, beginning with the New Encouragement of Investment Law in 1972, offered various incentives and protections for industry. The Central Bank of Jordan also initiated a series of policies designed to encourage commercial banks to lend more to industrial projects rather than to commercial enterprises.[15] Accordingly, the various economic plans issued by the government in the 1960s and 1970s routinely called for increased growth in the industrial sector over the commercial.[16]

Remittances also fueled an economic boom in Jordan, and the increased cash in the economy fueled increased domestic consumption. The resulting economic boom was expressed mainly in real-estate speculation, construction, and services. Amman became a modern city. The new domestic market also generated demand for cheap, mid-ranged, durable goods that could be produced in Jordan. Assisted by government incentives a small industrial sector did begin to develop. It would be an overstatement to suggest Jordan industrialized in the classic sense; however, small and medium-sized industries did begin to take root in the 1970s. This marked an important change within Jordan's business community. Where previously merchant interests focused solely on issues of trade and import, the seeds of conflicting interests began to develop. In 1961 industrialists within the ACC broke with Chamber leadership to form the Amman Chamber of Industry (ACI).[17] This move foreshadowed coming organizational problems among business elites centered in Amman. Moreover, it was symptomatic of the growing divide within the business community as new sectors of economic activity evolved. Broad business community concerns were no longer axiomatically tied to elite merchant interests. By the end of the 1970s the trappings of the semi-rentier state and rentier economy in Jordan were quite evident.

A fiscally independent state and a more or less dependent private sector gave rise to "the unwritten agreement" between merchants and rulers. Similar to many other merchant–ruler relationships in the Arab World,[18] the funds available to the state for buying off some merchant elites carried a price tag. Merchant elites essentially opted out of playing any political role in the country. Merchants pursued profits from state patronage and refrained from the type of overt political advocacy that would endanger it. It was no surprise then that business realized its lowest level of policy influence in this period. Policy initiatives by the largest business representative, the ACC, went wholly unaddressed by the state. For example, in 1974 without consulting business elites, state officials created the Ministry of Supplies (MOS). This ministry implemented government subsidization schemes and actually took over the importation and distribution of meat, rice, sugar, tea, flour, cooking fat, and some fuel products. For other categories of goods the MOS established set prices. In other words, the MOS became a direct competitor to the private sector. Protests by the ACC to curb the power of the MOS or allow merchant representation within the ministry were refused by state officials.[19] Antagonism between business and state went so far that by 1979 "hundreds of merchants" were tried before an MOS-run court for price violations and hoarding.[20]

All of this was set to change by the early 1980s. The economic boom of the 1970s was virtually reversed in the 1980s. As early as 1982 it was apparent that foreign aid was declining. As oil prices in the region similarly declined, worker remittances dropped off. In 1981 grants (that is foreign aid) comprised 40 percent of the government budget. By 1990 that percentage had declined to just 14. This reduction by more than half contrasted to a domestic tax revenue increase of about 58 percent. The remaining shortfall in revenue was financed principally by greater borrowing resulting in an 86 percent increase in external debt over the same period.[21] Regional economic downturns also lowered remittance levels. In just one year, 1986–7, trackable remittances fell by roughly 25 percent.[22]

Greater debt and flat remittances combined to undercut the government's long-standing policy of a strong dinar. By November 1988 the dinar lost two-thirds of its value against the dollar over only a few days.[23] The quickening economic decline fed the government's pronouncements of economic reform. As early as 1986 Prime Minister Zayd al-Rifai's new administration announced numerous reforms such as cuts in government ministry spending, new luxury taxes, and fee increases.[24] By 1989 the government entered into negotiations with the IMF for debt relief. In hand with these reforms was the beginning of the government's initiative to create a new relationship with the business community. Hints of this first came at the inauguration of the Rifai administration where King Hussein called for "steps toward reviving and reinvigorating the economic processes in Jordan and for a program geared toward stimulating the private sector."[25]

What followed was a rather haphazard process whereby ministerial and executive-level committees dealing with economic policy were either created or opened to include private sector representatives. The first and most important of these committees, the Economic Consultative Council (ECC), was established in 1986. Located within the Prime Minister's office, this council was to parallel the long-standing Economic Security Council. Unlike the Security Council the ECC permanently included business representatives. Other similar bodies were established or strengthened within the Ministries of Planning, Commerce and Industry, and Supply. While a far cry from the infrastructure of full-blown state corporatism, these bodies marked a shift in state strategies. Previously, the bulk of contacts between business and the state was individual and informal. Beginning in the middle of the 1980s the Jordanian state made a conscious effort to draw organized business into a limited and managed, quasi-corporate arrangement.

On the eve of political liberalization in 1989, economic shifts and changes in government policy suggested a state more vulnerable to policy input, particularly on economic issues. The openings developed before 1989 became even more important with the advent of formal political liberalization. By all accounts the scene was set for a new pact to be established between merchants and rulers in Jordan, a pact that could economically support the political hopes for liberalization.

Political liberalization and the new economic reality of the 1990s

In April 1989 riots broke out in several southern Jordanian cities, ostensibly in protest over government austerity measures. Political liberalization's advent shortly after was a direct state response. The concern in this section is to chart the dynamics of political liberalization as well as the economic and social changes in the 1990s that shaped the way the private sector attempted to negotiate aspects of liberalization. Later, three issues of business–state concern from this period are examined to highlight how these factors were actually at play.

Elections for Jordan's first parliament in 23 years marked the beginning of increased state openness to wider policy participation. For the representatives of Jordan's business community this meant increased opportunities to lobby the state on economic policy. Two aspects were viewed as most important: greater press freedom and the re-institution of parliament. Increased press freedom was the most visible and exciting aspect of Jordan's liberalization. Jordan joined the ranks of Yemen and Kuwait where a relatively free press often injects itself into domestic policy debates. As expected, business organizations and interests often utilized this avenue to present their ideas on current policy debates. Though none of the major dailies in Amman is controlled by pro-business interests, there are columnists who frequently present private sector interests in these debates. Interviews with Chamber of Commerce officials, coverage of major banks' year-end reports, and opinion pieces by leading merchants are examples of uses of the new press freedom. There is of course a flip side. Other opposition groups have also quickly and more effectively pursued press exposure. The Islamic Brotherhood in Jordan has proven itself a formidable foe to business interests through its own media outlets. Many of these other voices, for example, press claims that the business community in general is out to rob the state or impoverish working Jordanians. While press freedoms have certainly imparted new opportunities, the larger impact has been an increased competition to express one's position amidst many loud opinions. Reconvening parliament went some way in institutionalizing this competition.

The new parliament of November 1989 and the ones to follow did not differ in their power or purview from the parliaments of the 1960s. Consultation and approval are its primary functions. Legislation can only be introduced through the Prime Minister's office and parliament serves at the pleasure of the monarchy and can be disbanded at any time. Despite these limits the presence of parliament did begin to subject economic policy-making more and more to the quasi-party politics of Jordan. It also changed the lobbying methods of organized business. Previously elite businessmen utilized personal government contacts; business organizations and powerful merchant families directly pressed relevant ministers for larger concerns, and as a last resort on important issues, groups of merchants would approach the Crown Prince. The presence of parliament added a new and complicated venue of policy participation for the private sector. The Finance Committee[26] in parliament rapidly became the primary body before which business representatives would go to plead their case. This meant that even after interaction at the Prime Minister's level to influence draft legislation, business leaders could still go before parliament to press for last-minute amendments or changes not accepted in previous lobbying. Combined with the creation of various policy boards prior to 1989, it appeared parliament provided firm foundations for a greater private sector voice in economic policy. Observers and advocates of the growth of civil society in the Arab World often heralded such examples as the means toward greater democratization. With new levers of access, civil groups can now expand their space in society and more effectively express demands to political authority. Business, with its established associations and well-defined interests, appeared best prepared to fulfill this role. Two caveats should be considered.

First, parliaments empower different groups differently. As with greater press freedom, groups such as Islamists and tribalists have proven more apt at managing the politics of parliament. Issues even tangentially tied to public sector employment, finance sector reform, free trade, and privatization often spark these groups' opposition to private sector positions. Leadership of some business associations made several attempts in the 1990s to establish a pro-business party but were unable to agree on the basics.[27] While there are businessmen in parliament, there is no organized representation or party through which private sector interests can be articulated. Second, representatives of the private sector have had difficulty mastering the art of lobbying in parliament. The largest business association, the Amman Chamber of Commerce, did not establish its own understaffed research and studies department until 1985. Parliamentarians complain – and many businessmen admit – that business presentations before committees are often ill-prepared and lack political sensitivity to the concerns of parliamentarians.[28] In other words, they cannot make a pitch well. In the case of Jordan, poor business advocacy has been due in part to factors rooted in the associations themselves. To make this point clear, let us examine the effect political liberalization in general has had on organized business.

Especially since 1989, elections to all social associations in Jordan have become highly charged and receive much press coverage. Since its founding in 1923, the Amman Chamber of Commerce has held free elections every four years for its executive board. Before liberalization these elections were viewed as indicators of changing business elites by the public. But after liberalization the public and the government have viewed these elections as bellwethers of municipal and parliamentary elections. For instance, the 1994 Amman Chamber elections and the 1995 Zarqa Chamber elections were quite heated. In Amman the incumbent president Mohammed Asfour was defeated by a relative outsider who effectively rallied the small business vote as a strike against larger merchant control.[29] The Zarqa elections were the first ever held by that Chamber and pitted a long-standing president against a strong slate of Islamic businessmen. After losing, the Islamic candidates unsuccessfully challenged the outcome. Both elections, but especially the Amman election, were symptomatic of the negative effect increased electoral competition has had on private sector representation. Instead of strengthening the position of organized business *vis-à-vis* the state, the highly contested elections in Amman have helped to shatter elite cohesion at the Chamber of Commerce. Slates of candidates are chosen less for their business profile or lobbying skills and more for considerations of securing large blocks of a diverse small business community.[30] Business representation has suffered.

Government officials often comment that business representatives "act as individuals rather than as representatives of the business community,"[31] and that they "present ill-defined proposals, or are simply not well organized."[32] Not only does this limit the level and quality of private sector input into economic reform but on issues of common concern to both the private and public sectors (foreign investment, for example), organized business in Jordan has proven a poor partner for the state. Three issues in the post-1989 period are discussed which bear this out: privatization, the 1995 Amman Economic Summit, and peace with Israel.

Privatization as a measure of business–state relations

As with IMF-sponsored structural adjustments in many countries, Jordan's privatization program was formulated in response to IMF recommendations on lowering the budget deficit and increasing revenue.

Privatization and its progress are integral measures of relations between a state and its private sector. The domestic business climate determines how attractive potential investments will be, especially to foreign investors. What the government sells, how it sells, and to whom it sells say a lot about how much political trust states elites have in the private sector. The progress of this process also suggests how much power a state is willing to relinquish to economic agents in society. Consequently, there is much more to the fate of privatization than simply abiding by IMF recommendations.

There are two important characteristics of Jordan's privatization program. First, the program is really a liquidation program of public assets. Privatization in the sense of turning over significant portions of public sector activities to the private sector is not on the table. Instead, a wide array of state investment in public shareholding companies is considered for sale. Actual privatization of wholly government-owned corporations (potash and phosphates for instance) or privatization of some government ministries is limited. In the case of the mining corporations the state has considered these strategic assets. Only the Telecommunications Corporation (TCC), the state-owned telephone company, has been seriously considered for sale. Progress on all these fronts, however, is slow. To date only government shares in the InterContinental Hotel in Amman have been sold. The state, instead of privatizing the TCC outright, has turned it into a public shareholding company (with all shares owned by the state) and allowed limited competition in services.

Second, the management and planning of the privatization program is also of note. Privatization in Jordan is less of a program and more of a direction. Various privatization documents have been issued with each government five-year plan, but there has been no comprehensive plan.[33] Consequently, the process is very decentralized. There is a technical committee for privatization located at the Prime Minister's Office which was created to oversee the privatization process. But, since most state assets in public shareholding companies are owned by the JIC (12 percent of capital on the Amman Financial Market, the local stock market), it is this body which generates many of the decisions of what to sell and to whom. Other shares are held by individual ministries; hence each minister has input into decisions over what to sell and to whom. Constraints on privatization reinforce these characteristics.

First, as previously discussed, state ownership in the past allowed state officials to reward and punish economic elites seeking positions on the executive boards of many public shareholding companies. As

Waterbury has pointed out, giving up this power is a political decision and a difficult one because it essentially goes against the logic of state investment in these corporations in the first place.[34] Second, since there has been no real planning to the privatization effort, the role of the private sector and its associations has been limited. Business leaders almost unanimously express ignorance over state intentions toward privatization except what is broadly recognized. Consequently, there is little private sector confidence in the process. Third, there is an underlying political calculus to privatization that is unique to Jordan. Given that the historical development of the private sector has been primarily Palestinian in origin, the politics of East Bank versus West Bank is of prime consideration. This divide is easily overstated, but nevertheless it should be taken into consideration. Thus, it is obvious that most of the potential buyers for government shares will be of Palestinian origin. Selling large amounts of public shares to portions of this group threatens to alienate the monarchy's support base among East Bank political elites. Hence the only sale so far, the InterContinental Hotel, was to a consortium of investors headed by a prominent East Bank notable. It was of note, however, that most of the investors were merchants of Palestinian origin, showing that when it comes to business there is room for collaboration. Finally, there are few really attractive opportunities for investment among what the government wants to sell. For example, since the mid-1980s the government has been touting the sale of the national airline, Royal Jordanian. However, because the airline runs very high debts and it only leases its airplanes, there is little capital a buyer can sell off to make the airline profitable. What is attractive for investment, the mining industries and the TCC for example, are not what the state wishes to sell.

The slow progress of privatization is therefore determined by many factors, and though it is an integral aspect of economic liberalization and potentially influences political liberalization, there is little interaction between the business community and the state on this issue. State strategy in this instance signals fundamental distrust in its own private sector. Other more specific issues between the business community and the state reveal some of the complexity of this relationship.

Trying to work together: the 1995 Amman Economic Summit

The 1995 Amman Economic Summit tested the extent to which each could coordinate toward a similar goal. The Summit brought together countries of the Middle East (including Israel) to discuss economic issues focusing on cooperation and investment. It provided a stage upon which Jordan could sell itself to private foreign investment and international lending bodies. Both state elites and the business community desired such investment and both attempted to coordinate a unified approach.[35] That attempt, however, proved difficult.

As part of the larger regional peace effort the Amman Summit was conceived of as a follow-up to the 1994 Casablanca Conference. Where the Casablanca Conference was a political gathering meant to lay the groundwork for Middle East economic cooperation, the Amman Summit was intended to provide the opportunity for actual deals to be made. In concrete terms the Summit was a mad dash by some Arab states to secure foreign (principally Western) investment in their failing economies. Peace with Israel, a well-known liberalization record, and its role as host of the conference, placed Jordan at the head the pack. Hopes and rhetoric ran quite high in Jordan that the summit would spark the awaited peace dividend.

Shortly after the Casablanca Conference, the Minister of Planning approached the Canadian Embassy to solicit assistance on the preparation of Jordan's investment strategy for the summit. The primary tool to lobby foreign capital was a projects' list prepared by the government touting joint-investment projects in Jordan. Under a project financed by the Canadian Embassy, foreign consultants were hired in December 1994 to assess this list. One of their conclusions concerned the need to involve Jordan's private sector in summit preparation. The consultants noted that a key element in any strategy to attract foreign investment is a stable investment climate. The most simple and direct means for the potential investor to assess this is to sample the local business community's opinions. It is after all the health of the private sector which suggests chances for successful investment. One American participant to the summit, commenting on the nearly US$530 billion in Arab assets kept outside of the region, put it this way, "If the private sector has all this cash, why doesn't anyone invest? If they think it's a risk, how can I get my company to take that risk?"[36] It was this predicament which became a key concern of the Ministry of Commerce and Industry and Prime Ministerial officials as they worked toward the summit. To address this issue the Canadians helped develop a list of private sector representatives to participate in meetings on the final preparation of the project's list.

This was an excellent test of state–business relations because a relatively objective agency was given purview to examine Jordan's business community and attempt to select its elite members to work with the state. Their first task was to contact Jordan's leading business associations for lists of potential representatives. Instead of receiving lists of "dynamic business leaders," the tardy response only contained "the names of family members of the organisations leaders." Moreover, the Canadians concluded that these formal representatives could not "deliver mobilization and that it would entangle the process in bureaucracy."[37] Consequently, of the 61 individuals invited to the first meeting in May 1995, two were government representatives and only seven represented business associations. The rest were businessmen (and one businesswoman) chosen through the Canadians' own research.[38]

The first meeting allowed the private sector to compile its own list of priorities for the summit. These priorities called for projects at the summit to be "attainable and sustainable." Many of the projects the state had originally proposed were large-scale investment tasks such as the Disi–Amman water pipeline and the Red–Dead Canal.[39] There was little private sector interest in such endeavors. Suggestions from the private sector focused instead on well-known sentiments to reform current economic regulations, re-educate civil servants, and involve the private sector to a greater degree in economic policy-making. Minister of Commerce and Industry, Ali Abu Ragheb, summed up the government response at the meeting: "With the increase in economic activity it is very tempting to maximize profits at the expense of the consumer and the environment. Such imbalances cannot be tolerated even as the historical tasks ahead of us."[40] Aside from an exchange of views the problem remained as to how the private sector should structure its involvement in the summit. Since business association involvement had been weak, who was to decide participation and organization?

Eventually, the Ministry of Commerce and Industry with final approval from the prime minister appointed a ten member board, the Private Sector Executive Committee (PSEC). These ten leading businessmen then organized eleven sectoral committees, appointed its members, and organized follow-up meetings. By most accounts the PSEC was effective. Raising their own funds, the PSEC produced several well-organized publications for the summit and, most importantly, interacted smoothly with state personnel organizing the summit. This interaction helped shape the final projects list which consisted of 27 ventures totaling roughly US$1.3 billion in private sector projects and US$3.5 billion in public sector projects.

The formation of the temporary PSEC was noteworthy for the lack of traditional business association involvement. Though the major business associations (Amman Chamber of Commerce, Chamber of Industry, and Jordanian Businessman's Association) each had a member on the PSEC board, wider representation and more significant involvement was not present. Several officials involved in the summit confirmed that the business associations were not supportive of the PSEC and viewed it as a rival to their own authority.

Though the Amman Summit proved an organizational success for Jordan's public sector, it demonstrated little in the way of a new pact between state and business. To an extent this outcome was to be expected since liberalization has brought little meaningful dialog, national or closed-doors, between the private sector and the government on the macroeconomic future of Jordan. Moreover, institutional weaknesses and incoherent leadership among private sector representatives doomed effective policy participation. There was no ready, strong partner with whom state ministries could coordinate, so state agencies were forced to create a temporary one. Shortly after the summit Dr Bassam Saket, Chairman of the Jordan Cement Factories, accurately described the situation, "Let us admit that the Jordanian private sector depends on the government to take the initiative. We have several private sector institutions but they are like heads without bodies."

The final issue shifts focus to state strategies and the business community's movement from quasi-participation to outright advocacy of a unique kind.

Jordan's peace with Israel and the business community's reaction

The economic benefits of peace with Israel have been an issue of significant concern in Jordan. For some it is the promise the Jordanian government can never keep. For others it is viewed as a calculated lie intended from the onset to convince a doubtful Jordanian public. In any case, the anticipated economic payoffs have been a concern for state elites, as failure to deliver could threaten the entire process. Shortly after the signing of the peace treaty, the Prime Minister's Office led repeated efforts to engage the assistance and participation of the private sector in working with Israel's private sector; if not in actuality then at least publicly, to demonstrate work toward realization of the peace dividend.[41] While many other professional associations took advantage of liberalization to voice their opposition, the business community adopted a low-key approach toward peace with Israel. This initial position was in line with the business community's traditional dependence on the state and avoidance of involvement in external political issues. So, enthusiastic support was not forthcoming but neither was outright rejection. In 1996 all of this came unraveled. After continued public discontent over the absent peace dividend, the opening of the tunnel near Al-Aqsa Mosque in Jerusalem, and clashes in Gaza, business associations came out officially against further normalization. To understand the progression and importance of these events it is first necessary to examine attitudes within the private sector concerning peace with Israel and its economic meaning.

For one portion of the private sector there is strong opposition to normalization and any commercial dealings with Israel. Some of the nationalist, larger merchants in the Jordanian Businessman's Association for instance share this view, as do some of the smaller merchants of Palestinian origin in the Amman Chamber of Commerce. A second segment is less ideologically concerned about the peace; rather they worry about economic dominance by Israel.[42] Since few of Jordan's industry and export sectors are prepared to compete for Israeli markets, this particular concern has been prevalent among Jordanian industrialists. A third, albeit quiet, opinion looks forward to trading and competition opportunities with Israel and the new Palestinian Authority. This belief is most prevalent among the largest traders and industrialists whose capital positions and foreign investments would hardly be injured by trade with Israel. Specific sectors of the economy, textiles for instance, would also naturally benefit from joint ventures with Israeli producers. Finally, it is no secret that segments of the private sector have long dealt with the Israelis through trade in the occupied territories. So it was in the face of these divided interests that the state sought to solicit assistance from business associations and leaders to increase support for normalization.

One of the traditional areas of close business–state cooperation has been in hosting foreign delegations to attract investment. Accordingly, in 1994 and 1995 the state quietly pursued the leading business organizations to participate in joint Palestinian–Israeli business meetings. One of these initiatives involved a three-day meeting in July 1995 with Israeli, Palestinian and Jordanian businessmen at the Royal Court.[43] Throughout this period only the Jordanian Businessman's Association, among the business associations, flatly rejected any participation, albeit quietly. The Amman Chamber of Commerce and the Chamber of Industry did participate in some gatherings, but kept the publicity low and had its officials play down the significance of the meetings.[44] The Amman Summit marked the high point of state efforts to generate more interaction between the two private sectors. At the summit, Egyptian, Palestinian, Jordanian and Israeli representatives committed to form the Regional Business Center (RBC), a cross-national Chamber of sorts to promote private sector involvement in securing the peace. Though a US-initiated idea the RBC received strong support from King Hussein and Crown Prince Hassan.[45] Indeed, following the Rabin assassination the King brought the leaders of Jordan's business associations with him to the state funeral.[46] None of these actions, however, generated much interest within the private sector. After a year of continued economic stagnation and setbacks in the progress of peace, business leaders began to reassess their low-key position.

Their first target was the easiest, the RBC. Beginning in October 1996, articles began appearing in Amman's dailies about the activities of the RBC. Leaders of all the business associations took the opportunity to come out strongly against any further dealings with the RBC because of Israeli actions in the occupied territories. Around the same time, leaders of all three business associations signed a declaration, along with opposition groups in parliament and professional associations, calling for an end to normalization and the withdrawal of Jordan's ambassador from Israel. This unprecedented political stance by Jordan's business associations was followed a couple of months later by protests over an Israeli trade fair in Jordan. Though business associations did not officially join the opposition organization created to fight the staging of the trade fair, the associations did openly side with the aims of the opposition. Business leaders and press accounts asserted these rare political stances were taken because of Israeli intransigence in the peace. While this may have served as an understandable justification, more instrumental reasons also influenced business's newfound political activism.

Problems with the peace process gave organized business the opening it wanted to bring pressure on a government that it believed had consistently blocked genuine policy participation over the past few years. Since the late 1980s, the unfulfilled promises of a greater role for the private sector in Jordan's economic policy and a decline in available state resources to buy off business discontent, grated on business elites. Business's overt stance against the peace was clearly an instrumental maneuver meant to be traded in the future to earn policy leverage. The row over peace thus reinforces two observations present in the earlier issues. First, the articulation of business interests through its organized associations and ability to sway state policy remain shallow in post-liberalization Jordan. The activism demonstrated by Jordan's private sector against peace, while unique historically, is far from the kind of sustained, sophisticated involvement required to support active participation in economic policy. Between policy participation and advocacy, Jordan's business associations fair better at lobbying but only when allies are present. However, much of that advocacy remains reactive and aimed at maintaining rent concessions from the state. Second, the row reflects that, even after almost a decade of rhetoric in favor of a new pact with the private sector, there has been little progress. Despite various policy boards and state initiatives toward the private sector, a partnership between the two has not developed.

Conclusions

Business–state relations in Jordan have not undergone the kind of changes that basic rentier theory predicts or prominent expectations of business–states in the developing world suggest. Instead, business's institutions of representation have failed to successfully advocate policy that would either endanger or advance Jordan's political liberalization. This basic lack of activism on the part of business has also limited the ability of state and business to create a pact in support of political and economic liberalization. Without business, economic reform in Jordan has been fitful for well over a decade. While there are other factors that also explain Jordan's limited political liberalization, the role of business is often overlooked. Aside from some temporary alliances with the opposition, Jordan's business community remains isolated from either an overt political role or a closer partnership with the state.

The expectation that, as the state retreats or creates new avenues of participation, social groups (especially business) will axiomatically gain a greater voice in policy requires more nuanced analysis. In Jordan, political and economic liberalization opened avenues for influencing policy not previously available. Instead of easing business efforts these opportunities complicated lobbying strategies and unleashed new pockets of resistance to private sector interests. Consequently, the evidence suggests that the impact of the private sector on democratization is very conditional. Against most expectations, liberalization in Jordan (both economic and political) is not dictated by the private sector.

So what does all this mean for the future of business–state relations in a hesitantly liberalizing Jordan? The prognosis is not good. Even before the death of King Hussein in 1999, Jordan's economy was showing signs of nearing crisis. In July 1998 Jordan reported to the

World Bank that its GDP growth figures for the mid-1990s were grossly over-reported. Instead, GDP growth has not outpaced Jordan's population growth (over 3 percent annually), an ominous sign. The new king, Abdullah, is unproven, most especially on economic issues. However, some short-term solutions are on the horizon. Increases in US aid and the promise of future Gulf aid may provide some relief and postpone serious restructuring of the economy. Likewise, such resources also ease pressures on the new king to move Jordan's political liberalization out of its deep freeze. But none of these is an enduring solution. The real problem is longer-term economic restructuring that will support the political hopes of any democratization. It will be exceedingly difficult for the state to carry this out without greater private sector involvement. Simply opening new policy venues does not deliver economic policy-making into the hands of the private sector, nor does it result in a new pact between rulers and business. More basic dynamics must be addressed for that to occur, and trends in that direction are not apparent in Jordan.

Notes

1 Support for my research in Jordan was kindly provided through two grants: an International Fellowship from the Social Science Research Council, Programme on the Near and Middle East, New York, NY (1996) and the American Center of Oriental Research, Amman (1994). Other funds were provided by the Inter-University Consortium for Arab Studies, Montreal, PQ. I would like to thank the following persons for their assistance and comments on earlier drafts of this chapter: Rex Brynen, Richard Doner, Riad al-Khouri, and Jillian Schwedler.

2 For example, Margaret Pearson, *China's New Business Elite: The Political Consequences of Economic Reform* (Berkeley, CA: California University Press, 1997). See also David G. Berger, "Business Associations in Latin America: The Venezuelan Case," *Comparative Political Studies*, 23, 1 (April 1990) 126–30.

3 The volume by Syliva Maxfield and Ben Ross Schneider, *Business and the State in Developing Countries* (Ithaca, NY: Cornell University Press, 1997), contains case studies demonstrating patterns of positive externalities throughout the developing world.

4 One of the best examples of a state–business pact that influenced solid economic performance is probably Chile. See Eduardo Silva, "From Dictatorship to Democracy: the Business–State Nexus in Chile's Economic Transformation, 1975–1994," *Comparative Politics* (April 1996).

5 Guiseppi DiPalma, *To Craft Democracies: An Essay on Democratic Transitions* (Berkeley, CA: University of California Press, 1990).

6 For an overview of rentier theory see Giacomo Luciani, "Allocative vs. Production States: a Theoretical Framework," in Giacomo Luciani (ed.), *The Arab State* (Berkeley, CA: University of California Press, 1990).

7 Jill Crystal, *Oil and Politics in the Gulf: Rulers and Merchants in Kuwait and Qatar* (Cambridge: Cambridge University Press, 1995); Kiren Aziz Chaudhry, *The Price of Wealth: Economies and Institutions in the Middle East* (Ithaca, NY: Cornell University Press, 1997).
8 See Abla Amawi, "The Consolidation of the Merchant Class in TransJordan during the Second World War," in Eugene Rogan and Tariq Tell (eds), *Village, Steppe, and State: The Social Origins of Modern Jordan* (New York: British Academic Press, 1996).
9 *Al-Iktisadi Al-Ourdoni, 1994* [The Jordanian Economy] (Amman: Annual Report of the Amman Chamber of Commerce, 1994).
10 Interview with Hamdi S. Tabbaa, Chairman of the Jordanian Businessman's Association, Amman, 27 June 1995.
11 Interview with Mohammed Tijani, former General Manager of the Amman Chamber of Commerce (ACC), Amman, 31 May 1995; also Zayd Sha'sha, "The Role of the Private Sector in Jordan's Economy," in Rodney Wilson (ed.), *Politics and Economy in Jordan* (New York: Routledge, 1991).
12 *Al-Kitab al-Dhahabi* [The Golden Book], Amman Chamber of Commerce, Amman.
13 *Yearly Statistical Series (1964–93)* (Amman: Department of Research and Studies, Central Bank of Jordan, October 1994).
14 Timothy J. Piro, "Privatisation in Jordan: the Political Economy of Public Sector Reform." Paper delivered to the Middle East Studies Association Conference, October 1992.
15 Michael P. Mazur, *Economic Growth and Development in Jordan* (Boulder, CO: Westview Press, 1979), p. 233.
16 *Plan for Economic and Social Development*, Years: *1964–70*; *1973–75*; *1976–80* (Amman: Ministry of Planning).
17 Interview with Ali T. Dajani, Advisor, Amman Chamber of Industry, Amman, 4 June 1995.
18 Crystal, in *Oil and Politics*, was the first to explicitly theorize about the unwritten agreement within the context of rentier state theory.
19 Interview Hassan, Tabba.
20 *MESC*, vol. 5 (1980–1981), p. 639, interview Tabba.
21 *Government Financial Statistics Yearbook* (*GFSY*) (Washington, DC, 1992).
22 *Economist Intelligence Unit Reports*, No. 3, Jordan (*EIU*) (1988), p. 9.
23 *New York Times*, 7 November 1988.
24 Ibid.; *Foreign Broadcast and Information Service, Near East and South Asia*, 23 January 1989; *EIU, Jordan*, No.1 (1989), p. 8.
25 Quoted in *EIU, Jordan*, No. 2 (1985), p. 17.
26 Given its purview over budgetary and tax legislation the Finance Committee has become the most important target for business lobbying in parliament.
27 Interviews with Mohammed Asfour, former President of the ACC, Amman, 24 May 1995, and Anis Muashir, General-Secretary Al-Wahd Party, 3 December 1996.
28 Interview with Sami Ghammo, former Minister of Finance, 3 July 1995, as well as impressions from businessmen who had gone before this committee and witnessed official business presentations.
29 Interviews with Tabba and Asfour.

30 Interview with Ahmed Obeidat, former Prime Minister, 5 June 1995.
31 Interviews with Gammoh, 1995; Dr Ziad Fariz, former Minister of Planning, Amman, 5 June 1995; Dr H. Khatib, former Minister of Energy, Amman, 6 June 1995.
32 *EIU, Jordan,* No. 3 (1989), p. 11
33 Interview with Bassam Saket, President, Jordan Cement Companies, December 1996, Amman.
34 John Waterbury, "Democracy without Democrats," in Ghassan Salamé (ed.), *Democracy without Democrats? The Renewal of Politics in the Muslim World* (London: I. B. Tauris, 1994), p. 215.
35 Interview with Daniel Joly, First Secretary, Embassy of Canada, Amman, 1 June 1995.
36 Quoted in *The Middle East* (December 1995), p. 17.
37 List of attendees provided by the Canadian Embassy, Amman.
38 *Investing in Jordan,* marketing documents for the Amman Economic Summit, prepared by the Investment Promotion Department, Ministry of Planning, Amman, 1995.
39 *Jordan Times*, 8 May 1995.
40 Interview with P. V. Vivekanand, editor, *Jordan Times*, 7 November 1996.
41 For example, some businessmen invited to participate in the Amman Summit preparation refused because of the involvement of Israel.
42 *Middle East Economic Digest* (21 April 1995), p. 10.
43 Interviews with Asfour and Amin Y. Husseini, Secretary-General, Federation of Jordanian Chambers of Commerce, 29 July 1995.
44 Interview with Bill Fisher, American representative to the RBC, Amman.
45 This was a rare event. The author found that only once previously had the king brought a representative of Jordan's business community along on a state visit.
46 *Al-Dustur*, 2 October 1996; *Al-Rai,* 2 and 9 October 1996.

9

Neoliberal Reforms and Democratization in Tunisia: Civil Society in a Rural Community

Stephen J. King

Introduction

The level of analysis in the literature on neoliberal transformation and democratization is usually at the national level. However, the impact of these policies on the political and social structure of a nation can perhaps best be discerned at the local level. In this chapter I offer this micro-level view. In particular I focus on the impact of neoliberal reforms on the elements of civil society that enhance the prospects for democratization. The community is Tebourba, a large village or town in Tunisia's rich northwest agricultural region.

I argue that neoliberal reforms favored large landowners at the expense of small and medium peasants. In order to avoid the social unrest this policy might have caused, state officials and the wealthy revived communitarian institutions emphasizing reciprocal moral obligations. The moral economy response consists of reviving both traditional patronage and kinship networks and explicitly Islamic values and institutions; to head off instability, the bureaucratic establishment in Tebourba encouraged the cleansing of wealth via the distribution of aid on religious holidays in an explicitly Islamic hierarchy. As a consequence, associational life in Tebourba appears to be heading in a direction that probably does not enhance the prospects for democracy. Associations based on blood ties and religion are promoted. While primordial identification increases, membership in organizations representing cross-cutting cleavages has declined during economic reform; a formerly active branch of the farmers' union in the community has stopped meeting altogether because it is now perceived strictly as the domain of the wealthy. Participation in other formal institutions is also on the decline. In sum, civil society in Tebourba has become less vibrant during state-led economic liberalization, and the forms of civil society thought to be the most helpful for democratization suffer while potentially detrimental forms are promoted.

This argument will be developed in the next three sections. First, I will discuss civil society and its importance for democratization. In the second section I will focus on neoliberal reforms and civil society in Tebourba. In the conclusion I synthesize the argument and suggest how these policies could have been implemented in a manner more conducive to a vibrant civil society and democratization.

Civil society and democracy

According to most theorists, neoliberal reforms enhance the prospects for democratization (if they do in fact lead to economic growth).[1] This literature builds on Seymour Martin Lipset's argument that democracy is related to the state of economic development. The more well-to-do a nation, the greater the chances that it will sustain democracy.[2] Lipset advanced several hypotheses about the causal dynamics involved: economic development promotes democracy by generating more democratic values and attitudes (possessed particularly by a more autonomous bourgeoisie), a less polarized class structure, a larger middle class, and a more vigorous, autonomous civil society. While these variables are interrelated, let us focus on civil society and democracy.

If we define civil society simply as associational life, there is a substantial literature going back at least to de Tocqueville,[3] as suggested in the introduction of this book, supporting the hypothesis linking a vigorous civil society and stable democracy. De Tocqueville noted the mutually reinforcing relationship between participation in civil society and participation in political life, depicting associations as large free schools where political interests were stimulated and political and organizational skills enhanced.[4] Embedded in this view is the notion of the civility of civil society. Ideas of pluralism and tolerance become enhanced through civil society. In addition, Lipset noted that a vibrant civil society checks and balances the power of the state.[5]

It is hypothesized that economic development, especially through a market economy, generates a more vibrant civil society for several reasons: it physically concentrates people into more populous areas of residence while at the same time promoting cross-cutting cleavages; in addition, economic development disperses literacy, knowledge, income, and other organizational resources.[6] Thus the recent market reforms enhance the prospects for democracy in less developed countries by sustaining a complex societal division of labor, generating a variety of cross-cutting economic and other interests, and undercutting the ascriptive, primordial identifications of religion, ethnicity, tribe and clan.

Primordially based segments of civil society (depending on their purposes and ideologies) may oppose democracy or harm the prospects of democratic consolidation. These forms of voluntary associations may promote intolerance, divisions, and immoderate interactions. The argument that follows suggests that state-led economic liberalization in a rural community in Tunisia has promoted primordial identifications while hindering the development of civil society based on associations considered more likely to contribute to democracy.

Civil society in a hegemonic party state

In the four decades since Tunisian independence in 1956, a single political party has monopolized the political system and resisted relinquishing control of the state apparatus. During the drive for independence the party was named the Neo-Destour. Early in the 1960s the party name was changed to the Socialist Destour, reflecting a shift in ideological direction. In 1987 leadership of the party was transferred, through a bloody coup, from the head of the nationalist movement, Habib Bourguiba, to current president General Zine Al-Abidine Ben Ali. This began a period of accelerated economic liberalization. The party was reorganized and the name changed to the Democratic Constitutional Rally. It is commonly known as the RCD for the French acronym. In all of its incarnations the hegemonic party has been led by party leaders and the bureaucratic establishment, especially the Ministry of the Interior.

Tunisia has a dense network of secondary associations. In a country of less than 10 million people, according to the Ministry of the Interior there are 3171 cultural and artistic associations, 822 athletic associations, 509 charitable, relief and social associations, 400 friendship associations, 126 associations for development, 115 scientific associations, 41 associations of general character, and two women's associations.[7] There are also active trade unions in Tunisia that have moved in and out of close corporatist arrangements with the state over time. On the level of electoral politics, currently there are six legal opposition political parties. However, all forms of associational life in Tunisia including trade unions and political parties confront a hegemonic party that is determined to maintain virtually complete social control and political dominance.

The regime utilizes several strategies to prevent associational autonomy. First of all no associations can exist without approval from the Ministry of the Interior. Visas to form associations are thus granted or withheld in light of regime interests. The regime routinely denies visas to associations suspected of independent Islamist inspiration. A banned Islamist party, *Al Nahada*, poses the most significant challenge to the regime. In addition to the legal constraints of the Ministry of Interior's visa system, the regime also utilizes financial controls to contain associational autonomy; subsidies for associations are distributed on the basis of political subservience. RCD loyalists also attempt to infiltrate new associations and duplicate them by forming ones of their own. Finally, the regime has not been reluctant to use force to contain society.[8]

At the local level, political control is orchestrated through an alliance between large landowners and officials in the Ministry of the Interior. The delegué is the most powerful local official and he supervises district or neighborhood heads called *Umdas*. In sum, social control in city and countryside has been an important characteristic of Tunisia's authoritarian single-party regime since independence.

As the previous section discussed, many analysts hoped that market reforms would bolster civil society, increase its autonomy, and activate the forms of civil society most likely to enhance the prospects of democratization. Tunisia began a period of gradual economic liberalization in 1970. The major political issue in Tunisia in the 1970s and 1980s concerned the degree of economic liberalism: some people within and outside of the party fought the development of what they saw as unbridled private-sector capitalism; how much of the state-dominated economy should be privatized was hotly debated. In spite of the general liberal economic direction undertaken in the 1970s, the government's share in total capital investment never dropped below 50 percent throughout the decade.[9] In the early 1980s the state still seemed somewhat committed to public, cooperative, and private economic sectors.

The Tunisian state since independence had committed itself directly or indirectly to fostering industrialization, as well as to maintaining social programs such as consumer subsidies. Neither a state-led nor private-sector-led growth strategy was effective in preventing fiscal deficits and inflation while the ruling elites pursued their goals. Even during the period of gradual economic liberalization the Tunisian economy suffered from rising inflation, shortages of food and consumer goods, and the inefficient use of capital – much of it borrowed.

In the early 1980s the country experienced stagnant growth rates. Balance of payments difficulties led to borrowing and debt that reached a near-crisis level in the mid-1980s. Like many other countries Tunisia resorted to a short-term macroeconomic stabilization program and mid-term structural adjustment program in 1986 to gain access to finance capital that would relieve the crisis.

The national political trends associated with Tunisia's accelerated market reforms have been analyzed by the author elsewhere;[10] however, here I explore the impact of these policies on political and social structure at the agricultural-sector level; and within that at the community level in the northwestern town of Tebourba. Nothing I am aware of makes Tebourba odd or different, so I believe that trends in Tebourba reflect trends elsewhere in the Tunisian countryside.

Tebourba: market reforms and neo-traditional social organization

There is a jarring agrarian upper-class bias in Tunisian development policies supported by the World Bank since their first agricultural sector adjustment loan in 1986. Bank loans supported the distribution of state lands formerly held in state-managed cooperatives to large-scale farmers. These producers benefited from below market input prices and credit and technological support. Nearly 1 million hectares of state-owned crop land have been transferred "with at best neutral, and probably negative consequences for the poor. ... The government is expressly not distributing these lands to improve the land assets of the rural poor."[11]

Contrary to any hope for employment creation on the privatized farms, there has been no net job creation in agriculture under structural adjustment.[12] Typically the first step taken by the new managers of the former state lands is to dismiss large numbers of agricultural laborers.[13]

In general, large landowners have been the only real beneficiaries of structural adjustment. The export crops favored in the new policy environment require irrigation, refrigeration, and other resources that small farmers typically lack. The prices for rain-fed crops and livestock produced by the poor have actually dropped in this period.[14] Only modest to poor efforts to improve small farmers' access to credit and inputs have occurred in this period. Overall, the World Bank policies adopted in Tunisia support a regime determined to favor the interests of large landowners over those of the small peasantry. Interviews with rural Tunisians with insecure access to land and employment reveal their awareness, despondency, and occasional rage at state development policy:

> The workers have become beggars! Why? The sun shines on everyone. Normally the state looks after us all. Why give the land to the rich? They already have land. If you give them more they will no longer think of the poor. What are they going to do with more, buy another car? It's no good. You find people with 1000 hectares while others won't even have one hectare. The poor wanted land. Some farmers before got land and they're doing well. [In the early 1970s a small amount of state land was distributed to former cooperative workers.] ... If you have connections you can get land. ... Those who were fired like me always go to the administration asking for work. We tell them "you fired us, so give us something to buy bread". Nothing happens ....
>
> (Interview with the author in spring 1994)[15]

A second laborer:

> How could they give all that land to those people [the privatization of state-owned cooperatives carved out of the French colonial farms to wealthy farmers]. If I had one hectare, I could produce more than five or six hectares worth. I work summer and fall only. How can I live on six months of work. For example, I may work one week or two weeks with a cow owner until he sells the milk, and then I have to hit the road. He has twenty thousand dinars and I have ten dinars. I have a brain like you and him. Why is it that workers always stay workers? ... My boss may have thirty thousand dinars and when I ask for a loan I may get twenty dinars. Why should someone live below zero while others drive expensive cars. He uses you. He doesn't pay you on time. The owner's wealth, even his cars came from us. I have a brain and blood and I work, but I can't even buy boots for the winter's rain.[16]

Structural adjustment created considerable inequalities in rural Tunisia and the inequalities have political implications. In order to control social and political instability state elites fostered deliberate redistribution along neo-traditional lines. Field work carried out in Tebourba, during the implementation of neoliberal reforms in 1993–94, revealed these social and political trends.

First settled by farmers from Andulusia, the people of Tebourba have considered themselves part of a community for over 500 years. A community that has been intact as long as Tebourba develops an identity that can be revealed by examining the expectations that community members have of one another. In contemporary times there are still community norms that provide a set of expectations of behavior for the inhabitants of the town. The rules of behavior in Tebourba are sometimes ignored, but are followed widely enough to reveal themselves to the outside observer:

1. In desperate times one can turn to extended family members for help to survive.
2. The wealthy provide a range of goods and services to the poor: employment, gifts of grain and other agricultural products during the harvest and during the Islamic festivals of *Ramadan, al'Eid al-Kabir and al'Eid as-Saghir*; the use of their land for the grazing of the poor's livestock; and in the cases of loyal clients the extension of loans, and service as intermediaries in economic and bureaucratic transactions. The poorest of the poor are to receive alms and the *Zakat* (tithe of yearly income given to the poor) from the wealthy and middle-income.
3. The poor provide the wealthy with labor services at low costs, and public acceptance of prestige, and high social standing for the wealthy who follow rule number two.
4. A mental account of favors granted, whether large or small, is to be kept and reciprocated when possible. Related to this, community members are expected to continuously build up their connections and relations with people within and outside the community.

The traditional community norms, described above, evolved largely in a pre-market era when many peasants produced mainly for subsistence. The lack of mechanization in farming made the labor services of the poor more crucial then they are today. In addition to these circumstances, which no longer apply to Tebourba, over the years migration from other areas has altered the population and made it more heterogeneous. In spite of these social changes traditional social organization remains a viable option for many community residents and the actions of state agents play a strong role in determining whether or not social action moves toward the individualistic self-help market-oriented behavior which is presumably one of the goals of neoliberal reforms.

The social action of local political elites during economic reform is distinguished by their efforts to create more legitimacy for state policy by reinforcing the channels of a moral economy based on an Islamic hierarchy for allocating resources. The *umda*'s[17] themselves described the party's efforts in maintaining community welfare mechanisms through their coordination with wealthy landowners. This *umda'* was a former member of the National Guard:

> The poor are reluctant to ask for aid. The umda' meets with party members who try to figure out who needs help and what to present. Several sector heads will meet with their umda and during *Al'Eid al-Kabir, Al'Id as-Saghir,* Ramadan[18] and other holidays they distribute the aid. We get the assistance from the wealthy who are friends of the party. Some of the rich give directly to the poor. On religious holidays some people will give around thirteen dinars to a poor family or a sack of wheat. Since I've been here we've had good community solidarity. There was some trouble a year or so ago and they talked about building a National Guard Office across the street. However, after I worked together with the rich to help out the poor things quieted down. Just this year there were 80 gift packages given out during Ramadan in my sector.

Another *umda'*:

> During the harvest there is a lot of solidarity between the rich and poor. During Ramadan money is given out or oil, sugar, and wheat. Clothes and meat are given on Holidays. During *Al'Eid al-Kabir* lambs and goats are given to the poor. Party members go around to the rich and we see that sometimes a rich person will give a whole animal to a poor family. Many of them give the animals to me to give to the poor. Some give up to four sheep. *Hajj*[19] Bechir built an entire mosque. Three other large farmers bought a mosque. We have people who give land for a hospital.

A third *umda'*:

> There are strong connections in this sector. The rich think of the poor automatically. When times are hard, the state will meet with the wealthy and get assistance to distribute to the poor. During the first ten days of Ramadan and the last ten days the rich give to the poor. During *Al'Eid al-Kabir* lambs are given, as well as something

> during the small festival at the end of Ramadan. We get a lot of help for the poor. Party leaders have an account where people can give the *Zakat*. Many people give one tenth which we distribute to the poor. Ibn Hajoub left an inheritance from his olive oil enterprise for the poor. We gave twenty five portions of five liters each to the poor.

In addition to direct aid to the poor some *umda*'s hope that the privatization of cooperatives will lead to a reinvigoration of diffuse patron–client ties that were part of an earlier moral economy:

> The state failed with cooperatives because they poorly organized laborers and managers. The cooperatives are impersonal; they cannot help the poor in urgent situations like death in the family. The *Hajj* can help the people who work for him because he has a personal relationship. His workers are happier than when they worked on cooperatives. He helps them get their *awula*.[20] At harvest time, he sells wheat to his workers at wholesale prices. In addition, he always gives the *zakat* that helps with their provisions and he gives them loans. The Hajj doesn't have technical training, but he is a specialist at growing wheat. Even without irrigation, his wheat does well. At a glance, he can pick out the best quality seeds that people with degrees miss. His wheat is used as a model all over the country. That's why he was given the cooperative lands of wheat fields. The *Hajj*'s two sons run the place alone; they do the work of twenty administrators.

The general point is that local party officials have long involved themselves in traditional and non-traditional community allocation patterns. Presently state agents should be carrying out policies of economic liberalization. However, these agents are promoting redistributive activities and neo-traditional social organization. This is not a new phenomenon, as Eric Wolf pointed out, "Sharing of resources within communal organizations and reliance on ties with powerful patrons were recurrent ways in which peasants strove to reduce risks and to improve their stability, and both were frequently condoned and supported by the state."[21]

State policy in Tunisia reinforces attachment to traditional patronage and kinship networks and to Islamic values and institutions. However, market reforms are considered to be conducive to democracy because they undercut these primordial identifications and promote civil society based on cross-cutting economic cleavages and other interests. In addition, the participation of the majority of the peasantry in formal organizations has weakened during market reforms.

Norman Uphoff and Milton Esman have noted that farmers' associations in developing-country settings have a tendency to become the tool of wealthier farmers while poorer ones are out in the cold or forced to rely on traditional patron–client relations, "if such organisations become institutionalised, they would be as instruments of the large and middle farmers, while small farmers would be thrown back on traditional links of dependency on patrons or compelled to rely wholly on their own meager resources."[22] This pattern largely fits rural Tunisia; and other trends in formal institutions appear to be increasingly favoring the strong over the weak.

Wealthy farmers of Tebourba appear to be exercising their political power by not allowing the meeting of the farmers' union in recent years, thereby avoiding the voicing of disagreement over agricultural policies by small-scale farmers. When queried by the author large-scale farmers said that they did not have any place to meet. Poorer farmers claimed that the wealthy had many ways to keep them out of the union, including having a size of farm requirement; and at any rate the farmers' union was simply a tool for the rich to help themselves, so there was no reason for anyone else to attend. These circumstances vary from the situation described by Mira Zussman in her late 1970 study of Tebourba.[23] At that time the farmers' union had four functions:

> First and foremost – and in comparison with European and American unions – it is a forum for communication of Ministry of Agriculture objectives. Second, it is a setting for the small group of [Tebourba's] wealthy landowners. Third, it is a place of articulating problems and mobilizing farmers. And fourth, it is a place where poor farmers come to make contact with the wealthy – particularly when they need a favor.

There is no lack of desire in many segments of Tebourban society to change agricultural policies, yet the farmers' union did not meet even once in 1994 when the major policy changes began their full-scale implementation. I was informed that it had been several years since poor farmers even bothered to attend union meetings. Why the abandonment of even the small peasants' modest efforts to promote their interests via dependency links forged during union meetings? Why the lack of a forum to create the patina of egalitarianism between wealthy and poor farmers? One reason is that government policy has favored the wealthy so dramatically in recent years that poor farmers are not bothering with the charade. Also, wealthy farmers no longer need union meetings to promote their interests in agricultural policy: they have won those battles, and further meetings would only allow a hearing for the voices of the disgruntled losers of state policy.

Large farmers and local officials, often one and the same person, have also exercised their power to limit the formal participation of poor peasants in other areas of the country. Some of the handicaps to small farmers as they attempt to enter a market economy were documented by Richard Fraenkel and Mathew Shane in their study of Ebba Ksour in Northern Tunisia:

> Through elective office large farmers enjoy greater power in such participatory institutions as the local mutual credit union. Another restraint on small farmer access to the services of public institutions is the requirement that all applications for their services must be approved by the local officials of the Ministry of Agriculture. The Ministry of Interior is the most important organization determining the distribution of purchased resources for cereals production. ... The shaykh, the lowest official of the Ministry of Interior, is responsible for a territorial sub-unit of a delegation. The shaykh's approval is needed for applications made by farmers to be sent to the appropriate branch of a technical ministry.[24]

The shaykhs were recruited from among the local population. All of the shaykhs in the locality studied were large farmers who rented a high percentage of the land they operated from small farmers. Small farmers complained frequently of refusals by the shaykhs to accept their applications for institutional services.

After the reversal of the collectivization program the government recognized the need of helping the private cereals producers to start up production. Two policies were decided. The first was to make loans available to small farmers from the mutual credit union system. The second was to liberalize machinery credit policy. However, as we shall see, these policies did not permit the small farmer as an operator.

An examination of the amounts of loans made by the credit union after 1969 to small farmers showed that these were in fact consumption subsidies; it would not have been possible to start up cereals production even using the traditional technology with loans which were as small as 9 dinars. A second problem is the relative power of large and small farmers in the credit union of Ebba-Ksour. Lending decisions are made by the union's board, all of whose eight members are large farmers in the locality. This is not in itself evidence that small farmers do not get loans from the credit union, but examination of a list of loan recipients from 1970 to 1973 showed it to be generally true.

Fraenkel and Shane conclude that market conditions kept most small farmers in the traditional agricultural sector after they left cooperatives. Some rented their land to large landowners who could obtain the credit to mechanize and use high-yield seed varieties. Most of the dividends of modern technology went to large landowners. The rental arrangements and grazing terms for the livestock of the small farmers kept them tied to large farmers.[25]

Conclusion

Market reforms in Tunisia are not enhancing the prospects for democracy in that country. This community-level analysis sheds light on why political trends are not moving toward the expected liberalization. Market reform efforts have been too biased toward rural elites to promote democratization. Indeed, economic reform may be reinforcing authoritarian trends and moving communities away from a self-help market orientation. The state has reinforced primordial identifications that may encourage intolerance and divisions in social organization. At the same time state policy has alienated the majority of the community from the formal organizations that enhance political skills and educate the population in democracy. By way of conclusion, I will sketch out an alternative set of market-oriented policies that would have improved the chances of creating a liberal democratic political regime.

An ideal market reform program, one that improves growth, equity, and promotes democratization, likely follows a shared-growth approach.[26] In a shared-growth approach leaders establish the principle of shared growth, promising that as the economy expands all groups will benefit. First, leaders convince economic elites to support pro-growth policies. Then they persuade the elites to share the benefits of the growth with the middle class and poor. Finally, to win the cooperation of the middle class and poor, the leaders show them that they will benefit from future growth. Explicit mechanisms are used to demonstrate the intent that all will have a share in future wealth. These include comprehensive land reform, raising the prices of crops produced by the small peasantry, workers' cooperatives, some housing programs, and the establishment of programs to enhance small and medium enterprise.[27]

A shared-growth approach is conducive to democratization because it produces a less polarized class structure and a larger middle class. In essence this approach promotes the forms of civil society that are conducive to democracy. In time, the improved economic positions of various social groups will lead to a more vigorous, autonomous associational life and more democratic values and attitudes. Gaining economic power and political space aids the unions and working-class parties that will press for democracy.

However, the shared-growth approach to market reform is not commonly followed in less developed countries. Instead market-oriented policies follow what is commonly called a neoliberal strategy, or the Washington consensus, due to its backing by the international financial institutions located in Washington, DC. The neoliberal reform model sponsored by the IMF and the World Bank comprises stabilization and structural adjustment measures oriented toward trade liberalization, privatization, eliminating relative price distortions, encouraging export growth and agricultural production, and imposing austerity measures aimed at trimming governmental deficits.

Little in the neoliberal approach employs mechanisms to share growth during the market transition by targeting small and medium enterprises or aiding peasants and agricultural laborers with insufficient access to land and credit. The end-result is a bias toward elites in urban and rural sectors. Neoliberal reforms hinder democracy prospects by producing a more polarized class structure; by relying on a technocratic style of executive decrees to implement major policy changes; and by hindering the ability of political parties and trade unions to represent the interests of the working class and the peasantry. While policies have yet to change, recent publications by the World Bank acknowledge the potential benefits of a shared-growth approach and a need for a greater understanding of state-building and social organization in the countries undertaking stabilization and structural adjustment programs.

Notes

1 Several studies reveal the general optimism about market reforms leading to democracy. See Samuel Huntington, *The Third Wave: Democratization in the Late Twentieth Century* (Norman, OK: Oklahoma University Press, 1991); Larry Diamond, "Economic Development and Democracy Reconsidered," *American Behavioral Scientist,* **35**, 4 (March/June 1992), 450–99; and Carlos

Waisman, "Capitalism, the Market, and Democracy," *American Behavioral Scientist,* **35**, 4 (March/June 1992), 500–16. Several lines of arguments are used by scholars contending that processes of economic liberalization facilitate the emergence of democratic political regimes. The nucleus of this literature goes back to Martin Lipset's thesis that the organizational, social, and cultural prerequisites of democracy are largely a matter of the level of economic development. Thus, if economic liberalization creates higher economic growth then it aids democratization. In the process of economic growth a substantial educated middle class will develop demanding greater participation in politics; a bourgeoisie will grow to check state power; and the bigger pie will soften working-class opposition to capitalism. Martin Lipset, "Some Social Requisites of Democracy: Economic Development and Political Legitimacy," *American Political Science Review,* **53** (1959). Martin Lipset's work and the immense body of literature following it is reviewed in Gary Marks and Larry Diamond (eds), *Reexamining Democracy: Essays in Honor of Seymour Martin Lipset* (Newbury Park, CA: Sage, 1992). This view is shared by Harik and Sullivan in *Privatization and Liberalization in the Middle East* (Bloomington, IN: Indiana University Press, 1992). E. Huber, D. Rueschemeyer, and J. D. Stephens add that capitalist development lowers the power of the landlord class and raises the power and ability to organize the middle and working class, "The Impact of Economic Development on Democracy," *Journal of Economic Perspectives,* **7** (1993), 71–85. Recent economic literature based on an extension of the neoclassical growth model to include governmental functions and other elements and cross-national quantitative data indicates that Lipset may be right in the long term, even for countries not moving toward democracy presently. Robert Barro argues that "improvements in the standard of living – measured by a country's real per capita GDP, infant mortality rate, and male and female primary school attainment – substantially raise the probability that political institutions will become more democratic over time."See Robert J. Barro, "Democracy and Growth," *Journal of Economic Growth,* **1** (March, 1996), 1–27. It should be noted that Barro's conclusion touches on more than economic growth; the social indicators he includes point to a need to improve equity in order to promote democracy. Barro suggests that Western governments should propagate Western-style economic systems as the effective way to expand democracy in the world. The analysis offers little to help us understand how authoritarian regimes reorganize during economic reform even though authoritarian politics and liberal economics is a reality of political life in many countries for the indefinite future. Imposing market reforms by authoritarian means does not appear to raise issues for the author.

2 Quoted in Diamond, op. cit.

3 Alexis de Tocqueville, *Democracy in America, 1840–1956* (New York: Signet).

4 Diamond, op. cit., p. 123.

5 Ibid.

6 Ibid., pp. 124–5.

7 Eva Bellin, "Civil Society in Tunisia," in Richard Norton (ed.), *Civil Society in the Middle East* (New York: E. J. Brill, 1995).

8 Ibid.

9 Lisa Anderson, *The State and Social Transformation in Tunisia and Libya* (Princeton, NJ: Princeton University Press, 1986), p. 240.

10 See Stephen King, "Economic Reform and Tunisia's Hegemonic Party: the End of the Administrative Elite," *Arab Studies Quarterly*, **20**, 2 (Spring 1998).

11 World Bank, *Republic of Tunisia: Growth, Policies, and Poverty Alleviation* (Washington, DC: World Bank, 1995).

12 Ibid.

13 Stephen King, "The Politics of Market Reform in Rural Tunisia." Unpublished PhD dissertation, Princeton University, 1997.

14 World Bank, op. cit.

15 Stephen King, "Politics of Market Reform."

16 Ibid.

17 *Umda,* as suggested earlier, refers to sector heads of the community who are the lowest officials in the Ministry of the Interior which controls the countryside in Tunisia's authoritarian political system.

18 Important dates on the Islamic calendar.

19 Title of veneration referring to someone who has made the pilgrimage to Mecca.

20 Yearly provisions of staples.

21 Eric R. Wolf, *Peasant Wars in the Twentieth Century* (New York: Harper & Row, 1969).

22 Norman Uphoff and Milton Esman, *Local Organization for Rural Development: Analysis of Asian Experience* (Ithaca, NY: Rural Development Committee, Cornell University, 1974), p. 66.

23 M. Zussman, *Development and Disenchantment in Rural Tunisia: The Bourguiba Years* (Boulder, CO: Westview Press, 1992).

24 Richard Fraenkel and Mathew Shane, "Ebba Ksour," unpublished paper, p. 25.

25 Ibid., p. 25.

26 Adam Przeworski (ed.), *Sustainable Democracy* (Cambridge: Cambridge University Press, 1995).

27 World Bank, *The East Asian Miracle* (Oxford: Oxford University Press, 1993).

Part IV
Latin America

10
Economic Liberalization and Inequality in Mexico: Prospects for Democracy

Remonda Bensabat Kleinberg

Introduction

The changing role of the state in Mexico and its withdrawal from the economy under economic liberalization has led to a transformation in traditional state–society relations, and the erosion of the "social pact," a historical arrangement that allowed for co-optation, some distribution, some compromise, and social stability. It has been suggested that the resulting new economic relationships would lead to the dismantling of non-democratic or semi-democratic political arrangements. However, by shifting (as it did) the control and direction of the economy over to the private sector, the state's capacity to act in the interest of the "weaker" groups has diminished, risking coercion, and not co-optation or compromise as its main instrument of control. While Mexico's new economic model, based on deregulation, privatization, and the overall liberalization of the economy, was hailed as the solution to Mexico's economic and social woes, it has led instead to greater inequalities.

This chapter will argue that *political* liberalization has not been the immediate outcome of *economic* liberalization and greater private sector influence. Rather, economic liberalization and concomitant shifting alliances have created conditions, at least in the short term – for, not only greater inequalities, but greater political instability. Sectors – both business and non-business – otherwise dependent on the state for resources, are now losing access to the state and have few avenues open to them. This has made the transition to democracy, or at best the dismantling of corporatism, tenuous if not an event of the distant future. Indeed, powerful private sector interests are redefining corporatist linkages with the state and with labor in an effort to dominate economic policy-making decisions. The outcome after ten years of the experiment (including a free trade agreement) has instead been one of the worst economic crises since 1982, the increase of income inequalities and growing political unrest. Rather than undermining the economic model, it has strengthened the Ernesto Zedillo administration's (1994–2000) commitment to economic liberalization. The *Pacto* – or what is now called the *Allianza* – continues to make concessions to business over concessions to the labor and peasant sectors.

In order to demonstrate the effects of the new model and new political arrangements on different sectors, this chapter will examine the distribution of benefits in the capital and non-capital sectors, particularly between 1989 and 1996. Indeed, the extent to which actual benefits were concentrated was evident as early as 1985, when de la Madrid began implementing neoliberal economic policies.

Economic liberalization, strategic alliances, and inequality

After several decades of steady growth under heavily protectionist policies, increasing pressure from international financial institutions and sectors of domestic capital forced a change in Mexico's economic policy toward a more open, deregulated model. Economic liberalization under Miguel de La Madrid (1982–88) and more significantly under Carlos Salinas (1988–94), and later Ernesto Zedillo (1994–2000), included privatization of key state enterprises, amendments in foreign investment laws, changes in the banking system, and the total deregulation of the trade regime, the culmination of which was the ratification of the North American Free Trade Agreement with Canada and the US. As the Mexican state accelerated its withdrawal from the economy it effectively dismantled controls over the private sector and created the basis for a new alliance with capital.

Economic liberalization *per se* had no *absolute* effect on political liberalization. Political reforms were implemented as early as the Lopez Portillo administration (1976–82), with political party registrations made easier, and an expansion of seats to opposition parties. As has been evident, these reforms had little effect on the political stronghold of the *Partido Revolutionario Institucional* (PRI). Even as the state loses its primary role as "Rector of the Economy," the PRI, as we shall discuss below, continues to hold tenaciously and authoritatively to its traditional reins. Carlos Salinas managed to strengthen the executive during his administration, while doing little to increase the transparency of the political system. While there is some evidence that Ernesto

Zedillo's regime has allowed for political liberalization, the behemoth that has always represented the Mexican state has yet to redefine new political arrangements and civil society–state relationships. At the same time, the very nature of the economic changes had a profound effect on state–society relationships; a more open, transparent relationship ensued between the state and business. The strategic alliance which followed had an adverse effect on the traditional social pact between the state and the social sectors risking the erosion of corporatism and, worse, social stability.

The state, corporatism, and capital: post-revolutionary Mexico

The Mexican economy had suffered severe disruption from the years of social and political upheaval (1910–20). Infrastructure had been severely damaged, halting the growth process. By the final stages of the Mexican Revolution the administrations of Venustiano Carranza (1917–20), Alvaro Obregón (1920–24) and Plutarco Elias Calles (1924–28), were confronted with the need to reestablish (as well as create) the necessary conditions to develop the industrial and commercial sectors. While a powerful, albeit small, financial and manufacturing entrepreneurial class emerged from the rubble of the revolution after 1920, they were highly dependent on the local state for different forms of protection, particularly in the regionally based manufacturers of steel, cement, textiles, paper, beer, and glass. Outside these industrial groups, business was largely disorganized and not prepared to invest in the immediate aftermath of the revolution. State investment was thus essential for economic recovery, and it began to invest in areas facilitating and directing the development of the national economy.

While there existed 15 chambers representing commerce, industries and mines, they were weak and disorganized. In order to restructure and strengthen the private sector, Carranza initiated the formation of a national confederation of Commerce (CONCANACO) in 1917, and a National Confederation of Chambers of Industries (CONCAMIN) the following year.[1] While established through the state, these new business organizations were to act as autonomous institutions representing the interests of commerce and industry – at least until the consolidation of corporatism under Lázáro Cárdenas (1934–40).

The Constitution of 1917 had established labor legislation in Article 123, giving labor an eight-hour work day, the right to strike, the right to organize, safety and sanitary requirements and a number of other concessions, but it was not until 1931 that the Federal Labor Law was enacted. In the meantime, a Board of Arbitration and Conciliation consisting of labor, government, and business was established to settle disputes, and constitutional provisions gave wide powers to the state to intervene on behalf of labor. This did not bode well for industry, especially industry from the north of Mexico. The northern industrial state of Monterrey had well-entrenched company-run unions and its extremely powerful and influential elite had a great deal of control over both labor and the media. Growing state involvement would indeed diminish their control over the workplace.

It was within this context that the Employers Confederation of the Republic of Mexico (COPARMEX) was formed, destined to become what Carlos Ariola identified as "an association par excellence of the bourgeoisie," where a unified class could defend their own interests *vis-à-vis* labor and the state.[2]

Influencing policy in this early period of state formation in Mexico was a loose coalition that included a new elite of revolutionary generals, surviving members of the old landowning class, and powerful segments of the financial and industrial elite – particularly bankers and large-scale entrepreneurs.[3] Initially preponderant in the coalition was a powerful group of bankers which, alongside access to special concessions from the state, helped to influence the formation and reorganization of the economic and financial institutions of the Mexican state, namely the Central Bank and the Finance Ministry. Mexico's existing banking system had been geared toward serving as sources of short-term credit, and was as such ill-prepared to help finance the reconstruction effort.

Bankers and sectors of the industrial class who collaborated in financial reconstruction attained channels of influence in state policy-making via these two state institutions, particularly between 1917 and 1934. This sector would represent and support a neoliberal faction within the new corporatist state.

The first Mexican national political party was founded in 1929 by Plutarco Elias Calles (1924–28) and his supporters. Named the Revolutionary National Party (PNR), its inception gave stability to the forthcoming presidential election and, by bringing in the military, caudillos, and rival parties, the PNR could claim to represent the revolution. The "official" party would run the central government and control all the elections at both national and local levels.

But it was Lázáro Cárdenas who consolidated the corporatist party system and initiated a complex process of co-optation, selected reforms, and socialization on the national level. Making post-revolutionary Mexico one of the most stable countries in Latin America, Cárdenas incorporated all of social sectors into the National Revolutionary Party (PNR). His strategy, included the restructuring of the PNR into the Party of the Mexican Revolution (PRM), on the basis of labor, popular, agrarian, and military sectors. The military sector was formally eliminated in 1940 and became part of the popular sector after 1943. The PRM was renamed the Institutional Revolutionary Party (PRI) in 1946, the party that was to lead Mexico up to, and including, the present day. The basic structure of the PRI had not been altered until the 1992 political reforms.

Cárdenas offset patronage traditionally provided by landowners by institutionalizing government–peasant, patron–client relations. Landless "peons," sharecroppers, agricultural wage earners, owners of small land parcels, and recipients of government land grants, were all incorporated into the National Peasant Confederation (CNC) in 1935. Forming the basis for the labor sector of the official party and dominating labor sector politics, was the Mexican Workers Confederation (CTM) established in 1936. By integrating all the relevant sectors of Mexican society into the core of the official party, Cárdenas both strengthened the state and enabled sectors to pursue their own interests within state-controlled boundaries. The rationale was that the state had an unbiased conception of the common good and the general interest, and thus the transformation of the country into a socially just, modern industrialized society, could only be led by the state. What did result, however, was more of a corporate centralism, based upon the principle of corporate control and geared toward managing and channeling conflict through institutions.[4]

Both Cárdenas's strategy and logic had its implications for business–state relations. Concomitant to centralizing power on the basis of institutional arrangements, and achieving legitimacy of the state on the basis of its role as sole overseer of the economy, was decreasing the private sector's influence over the state. From the outset Cárdenas thus imposed stronger state control over capital, legitimizing Mexico's revolutionary myth, and giving labor an ally within the state.[5] Initially, control over the private sector came in the form of its actual exclusion from the official party. Within the corporatist structure established by Cárdenas, private sector relations also occurred through alternative, state-controlled channels: companies were now *legally required* to join the major business chambers, the Confederation of Chambers of Industry (CONCAMIN) and the Confederation of Chambers of

Commerce (CONCANACO). The states relationship with business had to essentially remain non-transparent; thus business was prohibited access to bureaucratic positions or posts in popular elections, and individual "political views" were effectively proscribed. This tended to both institutionalize the relationship between the state and the private sector, and reestablish the state's hegemony over the private sector. In this way the state remained in "control" over the direction of the economy up until the implementation of the new economic model based on trade liberalization and deregulation.

Personal ties and informal contacts between the state and individual entrepreneurs helped to bring private sector goals to the decision-making table. Thus, while the Mexican corporatist state was authoritarian in its nature and relegated business to more *quasi*-corporatist organizations, the private sector believed the state would continue to serve its interests. To be sure, the economic elite shared an interest in maintaining a corporatist, authoritarian structure, alongside a heavily protectionist strategy of import substitution industrialization (ISI), in effect between 1940 and 1970. Moreover, the state established its own development bank, Nacional Financiera (NAFINSA) which provided important credit to the new industrializing sectors.

The movement toward economic liberalization

State-directed growth within a context of semi-authoritarian corporatism allowed for two developments in Mexico: an annual growth rate of 6–7 percent and relative social and political stability for 40 years. Indeed, the enormous task of maintaining the delicate balance between capitalist interests and social peace without the use of overt repression was facilitated by periods of sustained economic growth.

That this would come to an end was clear by the late 1960s with the exhaustion of ISI and the government's repressive, fatal crackdown on a peaceful students' demonstration in Mexico city in 1968.[6] The financial and economic crises of 1976, however, and a subsequent agreement with the IMF for an Extended Fund Facility (EFF) set the stage for the liberalization of Mexico's economy. An oil boom in 1979 allowed for the essentially unpopular economic reform policies to be put aside. The period between 1968 and 1985 became one of increasing debate and conflict over economic and political policies, mitigated only during oil booms.

Though several measures were taken to open up both trade and domestic markets at this point, it was not until after 1985 – and the combination of a disastrous earthquake, an absolute drop in international oil prices, climbing interest rates and a collapse of the stock market – that the Mexican government accelerated its program of economic restructuring. The administration was again forced to sign an agreement with the IMF and World Bank. This time the international financial institutions had a significant say in the policy direction taken by the Mexican government. Alongside IMF loans, World Bank loans became more "policy-based" by the 1980s, conditioning their loans on deeper structural changes such as privatization, deregulation, the elimination of price controls and subsidies, opening up to foreign investment, and perhaps most relevant, relying on private sector investment to propel the economy.

Mexico showed its commitment to IMF policies by entering into the General Agreement on Tariffs and Trade (GATT) in 1986, arguing that immediate entry would control inflation and make industry more efficient. The fact was, however, that Mexcio's four largest sources of foreign exchange earnings (oil, maquiladoras, tourism, and migrant workers), were not covered by GATT disciplines, and membership into GATT threatened many of the less efficient industries and the associated labor unions that had prospered under the protective policies of ISI. Helped substantially by an unexpected fall in oil prices again in 1986, the ratification of the Protocol of Accession occurred, in the final analysis with very little debate. On 24 August 1986 Mexico became GATT's 92nd member.

Several critical trade liberalization measures were taken between May 1986 and October 1988. By 1987 Mexico had already exceeded the trade conditions attached to the World Bank policy loans, and continued thereafter to dismantle the system of protection. Privatization of more important state enterprises accelerated after 1986, and subsidies were reduced more than 80 percent in real terms on mass consumption goods sold at the state-subsidized marketing agency, CONOSUPO.[7]

The immediate outcome of economic restructuring has been the concentration of benefits and representation to big business over small and medium business, the weakening of labor in terms of bargaining power, wage and employment levels, and the increase in overall poverty in Mexico.

The concentration of benefits: the winners

At the earliest stages of economic liberalization, substantial gains went to big business. An extensive study conducted between 1984 and 1986 looked specifically at the "performance" of firms that participated in the Mexican stock market, and found that the size of business determined their success; the bigger the corporation, the more substantial their growth rates were during this period.[8] This regressive trend had much to do with the availability of credit, particularly after the formation of a parallel financial sector in the mid-1980s. The competition for bank credit between the nationalized banks and the stock-houses tended to drive nominal interest rates up, particularly since saving rates remained depressed. As a result, financing for small and medium businesses was difficult to obtain, and just as expensive as it was before the state nationalized the banking system. Access to the cheaper sources of financing through the stock market, moreover, was limited to only 200 corporations listed on the stock exchange.[9]

In 1985 alone the largest firms increased their sales by 50–100 percent, and obtained profits of between 100 and 700 percent relative to 1984. Thirty-eight percent of big firms and 42 percent of medium businesses reported similar profits, and only 25 percent of small firms recorded any major profits.[10] The same study also found an extremely high incidence of overlapping directorships, showing more of a trend toward economic concentration. Thirty-two businessmen, all of whom serve on the Board of Directors of 14 out of 19 of the largest conglomerates, participate in the committees and the Board of Directors of 59 of the major corporations registered. In other words, 10 percent of Mexico's leading entrepreneurs are involved in the corporate decision-making of 41 percent of the companies registered in the stock market.[11]

Moreover, these entrepreneurs figure quite prominently on the Boards and Councils of the particularly large conglomerates and more often than not, own significant shares in the business itself.[12] There is also a greater concentration of overlapping directorships in the big and "giant" conglomerates and interlocking participation on the powerful Executive Committees and Boards of Directors, has also opened up possibilities for further linkages between businessmen. This trend became stronger under the Carlos Salinas administration with the acceleration of privatization and the sale of former state-owned industries to big business.

While the intention of the Mexican state was to create a more deregulated and competitive economy, evidence exists that large companies have continued to remain the main benefactor. Big business in Mexico entered a new period of expansion after the first year of the Carlos Salinas administration. According to the major Mexican business journal, *Expansion*, Salina's economic strategy had everything to do with recovering business confidence and creating economic growth.[13] Overall profits of the largest corporations reached 6 percent in 1990, up 2 percent from 1989, and of the five businesses that reported the highest growth in sales, four were private domestic firms.[14]

Mexico's biggest industrialists have been restructuring, entering joint ventures, acquiring other firms, expanding into new markets, and investing heavily in their core sectors, since the economic crisis in 1982.[15] The commercial sector in Monterrey, as well as in the valley of Mexico and Guadalajara, saw its first major increase in ten years. In the valley of Mexico alone there were real increases in net sales of 71.5 percent,[16] and according to the National Securities Commission, by 1992 the Mexican security market hit an all-time high worth $US137 billion.[17] Big industrialists saw particularly large gains in sales and profits. Cementos Mexicanos (Cemex), owned by Lorenzo Zambrano, reported an increase in sales of 7.8 percent in real terms by 1992, with profits swelling by 30.22 percent. Cemex has continued to buy up other cement companies, making it the fourth-largest producer of cement in the world.[18] In fact, the Monterrey conglomerates, Vitro, Cydsa, Cemex, Alfa, and Visa, all owned by branches of two families (Zambrano and Garza Sada), have all reported increases in net sales and profits.[19] In 1991 the ten largest Monterrey groups (G-10)[20] invested up to US$2 billion, and achieved sales of US$10 billion, 5 percent of Mexico's GDP.[21] One of Vitro's subsidiaries in the United States acquired in 1989, Anchor Glass, increased its manufacturing alone by 4 percent, "resulting in the highest production in its history."[22]

Salinas also encouraged the formation of large holding companies, and authorized the establishment of financial groups which included bank and non-bank financial institutions.[23] According to the Mexican Bankers' Association (ABM), the formation of big financial groups was necessary in order to compete with foreign capital.[24] Over the span of one year alone, 10 out of the 13 financial groups authorized to operate in Mexico registered profits of almost 4 billion pesos between February 1991 and February 1992, and reported deposits of 17.3 billion pesos – equivalent to 50 percent of the total bank sales.[25] By 1992 large consortiums which included 15 groups had taken control of the financial system, with a total activity estimated at 350 billion pesos, equivalent to 35.6 percent of GDP, and more than triple the tax earnings expected in 1992.[26]

The big financial groups managed 93 percent of the capital of the banks in Mexico and 76 percent of the total capital in the entire financial system of the country.[27] Of these 15 authorized financial groups, only two – Banamex-Accival, owned by Hernandez and Harp Helu, and Bancomer,[28] owned by Garza Laguera – controlled 67 percent of the financial activity within Mexico, as well as 69 percent of investments in other countries. Together, the most dominant financial groups accounted for 71 percent of the profits by the close of the first trimester of 1992.[29]

The strategy of forming big economic groups in Mexico, combining financial, industrial, and commercial resources has led to what one Mexican businessman calls a "pyramid structure" of power.[30] The continuing trend toward joint ventures, the acquisition of more industries and financial institutions, and technology transfers, will tend to act to strengthen the pyramid structure, giving big business much more influence over economic policy.[31] This was clear during the free trade negotiations.

Big business and free trade

In a direct response to the state's request for private sector support and assistance in the free trade negotiations, the Mexican Business Coordinating Council for the Free Trade Agreement (COECE), was formed in September 1990. Headed up by Juan Gallardo Thurlow, co-chairman of the US–Mexico Committee of CEMAI, COECE organized itself into seven coordinating groups, six of which were sector-specific, and one dealing with cross-sectoral trade issues. Sectors, including the manufacturing industry, agriculture, finance/banking, securities, external trade, and commerce and services, all of which named a coordinator, were then divided into sub-sectors – 140 in all, of specific types of industry, commerce, and services. Coordinators were also appointed by the Chambers of each sub-sector to analyse the effects of free trade on each sector.[32]

After organizing the entire business class it was Gallardo's role to bring the appointed coordinators under one umbrella and to establish a common methodology for analyzing their individual sectors. Groups were then formed in each sector and sub-sector to analyse economic activity, competitiveness in the international markets, their position on free trade, demands and recommendations to be taken to both the Ministry of Commerce and Industrial Promotion (SECOFI), and the NAFTA negotiating table.[33]

By the first quarter of 1991 COECE had a general profile of the economic activities, positions, needs, and recommendations of each of the

140 sub-sectors. Position papers could now be presented to SECOFI through selected private sector negotiators. A complex set of meetings occurred during this period between the sectoral leaders, COECE coordinators, SECOFI officials, and the private sector negotiators – over 1400 meetings in all. Items were prioritized, then brought to the table, until a point at which the negotiators and the sectors held a joint position. Once agreed upon by the private sector and the state, the terms of negotiation were formulated and followed. Consultations between private sector representatives and the official Chief Negotiator, Herminio Blanco, extended well into the NAFTA talks; before each of the official negotiations the specific advisory group would have a meeting with the negotiator to discuss their terms.

While all sectors, and the main official and voluntary chambers, were directly involved in COECE, evidence exists that, even here, private sector representation was, in fact, highly concentrated. Seven of the big corporations belong to no less than seven chambers, with some belonging to 12 chambers, giving these individual conglomerates from 50 to 85.7 percent of private sector representation in the free trade negotiations. Not only were big corporations represented in several chambers that had a direct involvement in the COECE structure, but the sectoral coordinators themselves held senior positions in influential companies.

Indeed, while the whole structure of negotiations created "dialog" – in the form of regular, structured meetings, unprecedented in the private sector's relationship with the state, the dialog was clearly promoted largely between big-business representatives and the state. This alone obscured the fact that to a great extent the majority of the private sector *per se*, have limited resources to carry out time and money consuming debates. Indeed, several small and medium industries lacked any representation.[34] While it may be argued that big business has had some influence on government policy historically in Mexico, it is the nature of the relationship between the two sectors that reflects a new era for business. A decisive shift has occurred in the process of negotiation between the state and business.[35]

The corollary to the flourishing of big business groups under the policy of deregulation, privatization, and free trade, has been the alienation of small and medium industry. The tendency toward a more pyramidal structure has inevitably led to the diminishing capacity of small and medium business in directly participating in the new economic model. As an economist suggests, "there are two kinds of industry in Mexico, those that are booming because they capitalized their enterprises and are not debt-burdened, and those that go bankrupt."[36]

The fallout: the losers

Within the private sector the costs of economic liberalization were felt first by small and medium companies in terms of lost benefits, diminishing representation, and negotiating weakness. According to reports given by the main Mexican newspapers, one of the consequences of the liberalization of the market during the latter part of the 1980s was the disappearance of 28 000 micro and small industries.[37] The textile industry closed 1200 establishments since 1987, and has since experienced an annual decrease of 1.1 percent in the level of production.[38] Although capital has been invested in establishing industries, this has tended to occur alongside a significant loss of existing firms. According to the Chamber of Manufacturing Industries in Nueva Leon (CAINTRA), the industrial sector invested approximately 100 000 million pesos in 1992 in setting up 187 new businesses. While the new companies specializing in metal mechanics and fabrication of tools and equipment generated 1866 new employees, there was a simultaneous loss of 235 industries in the first four months of the year, with a loss of 1669 jobs.[39] What was difficult to determine was whether this loss in small and micro industries was a result of the economic crisis in 1986, or if it was indeed the effect of liberalization.

The trend toward a decline in small and medium enterprises was also felt in the commercial sector. According to the director of Pensamiento Empresarial de la Camara Nacional de Comercio de Monterrey, the monopolization of commercial "chains" has limited the possibilities for smaller establishments to compete in the market. It has also led to a series of bankruptcies. In 1991 the number of small and medium commercial enterprises decreased by 36.6 percent, alongside a 28.6 percent increase in large companies.[40]

Although there has been some attempt to redress the imbalance between big business and small and medium industries, the initiative has not come from the most powerful groups within the state. The Commission of Patrimony and Industrial Promotion of the Deputy of Chambers, presented a new scheme to be operated through NAFINSA. Within this scheme, referred to as "true popular capitalism," 120 000 small businesses were priced in the stock market, "once they demonstrate(d) a capacity to compete in the market." A budget of 18.2 billion pesos was also designated to small and medium businesses and oriented toward "endorsing the incorporation of Mexican business in the international market."[41] However, a study conducted by Instituto Proposiciones Estrategicas (IPE) in 1992, showed that, for every 1000 micro and small industries that emerged, 750 went bankrupt. One of the chief causes, according to IPE, was the lack of available credit.[42] NAFINSA director Oscar Espinosa concurred, but also suggested that the problems facing small business also included the lack of access to technical assistance, technology, and professional services, required to reach the level of development needed to confront international competition.[43]

The Chamber of Transformation Industries (CANACINTRA), also agreed that small business has lacked access to technical help. According to the chamber's own study the delayed incorporation of new technology to the production line has remained one of the most important obstacles to recovering the small-scale manufacturing sector. Alongside other obstacles affecting primarily micro and small industry, CANACINTRA included: the lack of technical consultation, little or no help in determining the quantity and type of scientific and technological development feasible for the individual enterprise, and the scarcity of capital to buy modern machinery and equipment. According to CANACINTRA the dire situation facing small business has impeded this sector from obtaining any adequate standards of quality and production necessary to compete in an open economy,[44] suggesting that "true popular capitalism" may be quite unattainable within the new economic model.

Indeed, the economic crisis has only served to worsen the situation for small business under the Ernesto Zedillo administration (1994–2000). In Monterrey, the business hub of Mexico, micro, small and medium industries are disappearing. The sentiment of the small business community is that they remain under-represented, and continue to feel the weight of the crisis, the cost of the free trade agreement, and the increasing competitiveness from external capital.[45] At the beginning of 1990 there were 25 000 small business members of CANACO–Monterrey, by mid-1996 there were only 16 000, 5000 of which are actually paying their dues.[46] In the valley of Mexico many small and micro retailers have also gone bankrupt. In June 1996 there were 25 000 members of CANACO–Mexico, down 10 000 from 1995.[47] The disappearance of small, micro and medium industries has had definitive repercussions on total employment. Small and medium business make up approximately 98 percent of the productive sector in Mexico, and employ over 60 percent of the economically productive sectors. Hence, unemployment figures have increased in most sectors (save the export sector) in Mexico.

Within the private sector the gap grows between small and big business, and their accessibility to resources, credit, technology, and international markets. While transitions occur within the private sector, the Mexican state's diminishing capacity to intervene in the interests of labor and the social sectors has only weakened the position of labor despite its involvement in the *Pacto,* or *Allianza,*[48] and despite the implementation of poverty-alleviation program.

Labor

Labor has weakened, in terms of bargaining power, wage levels, and employment levels. Under Zedillo the number of strikes has diminished. Hard data reveal the losses taken by the labor sector – in short- and long-term employment, in salary, and in buying power. A substantial shift had already occurred in the state–labor alliance, where labor had lost significant bargaining power *vis-à-vis* the state. Indeed, labor was one of the main losers in the shift in government economic policies.[49] Between 1982 and 1987 wages dropped by almost 30 percent. According to a study conducted by Robert Kaufman, wage earners' share of national income distribution in Mexico fell from 42.6 to 32.5 percent between 1981 and 1984 alone.[50] Short- and long-term unemployment increased between the first trimester of 1987 and the third trimester of 1995; where it was high to begin with, long-term unemployment alone increased by 21.9 percent. According to the Mexican statistical agency, INEGI, there was up to a 44.5 percent decrease of workers in the industrial sector alone. Labor lost significantly in several industrial sectors including basic metallics (44.5 percent), textiles (37.0 percent), wood products (32.9 percent), and the chemical industry (21.1 percent).[51]

Among the employed, both man-hours and salaries have decreased significantly. Man-hours dropped about 41.7 percent in the wood products industry, and 39.1 percent in textiles. Apart from the food, alcohol, and tobacco industries, man-hours worked seem to have decreased steadily between 1987 and 1996. Salaries took an even larger cut in the textile and wood products industries. Wage earners in the textile industry had their salaries cut by 51.8 percent in real terms from 1987 levels.[52]

To a great extent, however, the drastic peso devaluation in 1995, and the economic crisis that ensued, appear responsible for this drop. However, even before the economic crisis, where wage earners saw their salaries cut in half, workers' salaries lost 36 percent of their purchasing power between 1984 and 1992.[53] While only a "thumbnail" analysis was done on the figures, what seems to be clear is that labor is losing ground, both as part of the economically active sector (outside of exports), and *vis-à-vis* bargaining power. A Supreme Court decision was handed down in 1997 which dealt specifically with the public sector unions. The ruling allowed for more than one union in that sector – breaking down the stronghold of the Mexican Workers Confederation (CTM), over public sector unions. This vertically ties labor to independent industry eroding the CTM's horizontal linkages from industry to industry – at least within the state-owned companies which still exist. While the private sector unions are still dominated by the CTM, this could become a precedent for reversing control there as well, changing the latter's dominance over the Mexican working class more than any other ruling since the inception of Article 123 of the Constitution – the Federal Labor Law. The implication this ruling has on weakening the national labor union is clear. While entrepreneurs do not want to see unions dismantled,[54] precisely because organized labor has provided a basis for negotiating concessions to the private sector), business is promoting a deconcentration of power away from the CTM and into independent, industry-linked unions.

The weakening of the workforce as a result of the new economic model, as well as the economic crisis, has already been felt by labor in terms of their diminishing bargaining position within the new government–business–labor pact, the *Allianza*. It no longer has any "meaningful" access to policy formation, given the dominance over economic policy by the business class.[55] Concessions are rarely, if ever, given to labor as a result of *Allianza* meetings, precisely because its goals are rooted in anti-inflationary, orthodox measures and structural adjustment policies. Indeed, while Zedillo wants to raise wages by 17 percent, business is fighting to tie salary increases to productivity, a plan which is still under negotiation between the Businessmen's Employers Association (COPARMEX) president – Carlos Abascal – and the labor unions. Actions show that the nature of Mexican labor – without dismantling the fundamental corporatist structure – is undergoing deep changes.

The losses to labor and the drop in the value of social services by 23 percent have had implications for the increase in poverty in Mexico.

Poverty

While poverty figures in Mexico are not up to date, indeed the latest data on income distribution are for 1994, poverty and inequality have been growing very quickly in Mexico,[56] particularly given the present economic crisis. The poverty line was officially fixed in 1989 at two minimum wages per family, or a monthly income of 600 pesos (US$80).

Based on this measurement there are an estimated 42 million poor people, 22 million of whom are in the extreme poverty sector. Figures for the past eight to ten years, however, show an increase in poverty. Between 1984 and 1989 those in "extreme poverty," in mostly urban sectors, increased in absolute numbers from 4.3 to 6.5 million people, and declined to 4.8 million by 1992, still above 1984 figures. While this decline occurred in the urban centers, absolute numbers of those in extreme poverty continued to grow in the rural areas, both before and after the implementation of National Solidarity (PRONASOL).[57] The number of "indigents" increased by 1.7 million between 1984 and 1989 and increased by another 400 000 between 1989 and 1992.[58] According to Jonathan Fox, even with economic growth, extreme poverty will not diminish in the countryside unless the "Mexican government pursues an anti-poverty strategy that is much more targeted to employment creation and basic service provision in rural areas."[59]

According to Mexican economist Julio Bolvinik, while the intention of PRONASOL was to cushion social costs of economic restructuring, the poverty alleviation program (National Solidarity), introduced by the Salinas administration, made little difference to monetary income; it dealt specifically with social infrastructure.[60] In fact, the amount spent on the National Solidarity program under Salinas between 1989 and 1991 was less overall than the amount spent between 1980 and 1982.

PRONASOL was replaced by a more decentralized system under Zedillo, referred to as *Allianza Para Bien Estar.* Here funds are to be decentralized to the state and municipal governments. Spending, however, will still be determined by the federal government. The program combines the size of the population with poverty figures in that state – that is income level and unmet basic needs. Based on these figures the federal government will allocate resources to each state, with the latter determining the method for distribution to the municipalities. According to Bolvinik this has already created political conflict between PRI governors and PAN mayors in both Yucatan and Puebla, where the former have withheld funds from the latter.[61] The PAN mayors are fighting to have municipal funds put into the federal budget and made obligatory.

According to Bolvinik, the official economic policy – neoliberalism – is to blame for rising inequality. Poverty grew during the height of the Salinas Administration's economic program and the rise of income concentration. Bolvinik goes so far as referring to the model as a system of "Apartheid:" "only 20 or 30 million Mexicans are needed to run the economy well, the rest are disposable."[62] Indeed, to Bolvinik the economic model has failed: the economy has stagnated since 1981, and inequality grows. The withdrawal of the state from the economy diminishes its capacity to distribute resources. Any successful effort to reduce extreme poverty "will require some government intervention to redistribute income."[63]

The transformation in the nature of business–state relations, and the preponderance of big business in important policy-making forums reflect how far a new ideology has been introduced and adopted by the economic and political elite. Historically, the state almost appeared ashamed of its capitalist image and rarely dealt with business openly.

Economic liberalization: a path to democratization or greater control?

Without resolving the fundamental issue of "inequality" within the country, the new alliance between big business and the state risks creating increasing tensions and contradictions. Social turmoil, an element rarely experienced in Mexico since the revolution of 1910 is becoming more commonplace, particularly in the South – the poorest region of Mexico. The Chiapas-based Zapatista Army of National Liberation (EZLN) made its debut on 1 January 1994, immediately after the ratification of the Free Trade Agreement, demanding that the state address the deepening poverty and inequality issues in the South of Mexico. Indeed, on its heels came a variety of "people's organizations", including the People's Revolutionary Army (EPR) calling for the overthrow of the government altogether. The scenario is reminiscent of Latin America's southern cone in the 1970s, before the installation of ruthless, repressive, military regimes. Mexico's own, semi-authoritarian corporatist system was considered a model of stability that did not require a military government to quell civil society. Rather, co-optation, compromise, and some force (strategies lost with the erosion of traditional corporatism), helped to sustain some stability.

What rampant social disarray does suggest at the simplest level is that the PRI is losing its banner more and more, and without traditional linkages legitimizing the "revolutionary" party, the state may resort more and more to coercion in order to "resolve" increasing social instability before it turns into political and social crises.

In what direction, thus, is political liberalization moving? Several would argue that, on the basis of state–society transformations alone, a forum has been created for societal debate toward more open political institutions. That is – the argument goes – the redefinition of state–society relationships under economic liberalization will ultimately lead to democratization. Others, such as "transitologist" Phillippe Schmitter, suggest that democracy is not an inevitable outcome. In fact, as Schmitter states, "democracy is not (even) necessary: it does not fulfil a functional requisite for capitalism."[64] Nor, as the Mexican case demonstrates, was democracy necessary in order to implement a more deregulated capitalism. It was under a semi-authoritarian corporatist regime that economic liberalization was made policy.

New federalism versus old corporatism

To be sure, a process of political "liberalization" and state modernization has occurred since 1977 with the broadening of political channels to opposition party politics, and intensified in the 1980s and 1990s. Initially, administrative and political reforms were introduced to restore confidence and legitimacy in the state. The stipulation to legal registration, however, was that parties could not challenge the system itself.

Nevertheless, alongside changes in the electoral system, a "new federalism" has been implemented which purports to hand back power to the municipality.[65] Part of the challenge of new federalism has been redefining local power–civil society relations in an area rampant with patron–client relationships, powerful local bosses *(caciques)*, and heavily indigenous-populated communities and their concomitant traditional practises for choosing leadership. According to Luis Alfonso Carrillo's extensive, on-going study, indigenous communities continue to practice what he calls "archaic forms of democracy." Carrillo states that "of the 2393 municipalities in Mexico, 20 percent have traditional indigenous governments, electing their authorities on the basis of common law and precedent, and not through an electoral system. This is even more accentuated in some states, as for example Oxaca, where 80 percent of over 500 municipalities are indigenous and name their authorities in open assemblies."[66]

The imposition of modern apparatuses on traditional local culture and practises has created a great deal of tension. While decentralization has increased the number of state agencies and programs operating at the local level, resources, and the decisions around the distribution of resources, remain in the hands of either federal or state governments. Moreover, the increase of actors at both the state (municipal and federal), and civil society level has lent itself to the polarization of interests: old corporatist versus new ones, rather than to the forging of multiple pressure groups.[67]

Indeed, old corporatist interests dominated all federal "appointments." It is no secret that the "electoral" process that brought Salinas into power was exceedingly corrupt. Moreover, while liberalizing the trade regime through an agreement with its democratic partners to the north, Salinas (whose political cabinet were *politicos*) was concentrating power more in the executive. Salinas himself has argued that the simultaneous implementation of both economic and political liberalization policies remain incompatible.

It has been suggested that the extent and pace of economic liberalization, particularly under Carlos Salinas, would not have occurred *without* a heavy reliance on the extensive authoritarian powers of the executive. Roderic Camp states that, while there has been a decline in congressionally initiated legislation since the Echeverria administration, the near-abuse of executive powers has been observed since 1988. In the first year of the Salinas administration only 9 percent of federal legislation enacted was proposed by Congress. Moreover, tolerance of any opposition during this time actually decreased. More state governors were removed in the first two years of the Salinas administration than any other since the Miguel Aleman administration. And at the same time that the role of the state was being shifted away from the economy, Salinas also took measures to empower the executive.[68]

The very same regime successfully forged a new alliance with the private sector and guaranteed the irreversibility of economic deregulation through a North American Free Trade Agreement. Democratization would serve essentially a social "value" and may even threaten the basis of economic liberalization. The growing weakness of the social sectors – as well as small business – in Mexico, has eroded support for the free trade "model" and hence support for political parties that adhere to a mandate of economic orthodoxy. This was demonstrated in the 6 July 1997 Mayoral Elections. Leader of the left Party of the Democratic Revolution (PRD) Chautémoc Cárdenas became the first non-PRI Mayor of Mexico City – a "coup" in essence for the leftist opposition.

While entrepreneurs are now involved in the political "process" as both PRI and PAN candidates, the most powerful business interests support the PRI. Interestingly, rather than pursue the dismantling of involuntary "corporatist" associations, business is actually encouraging the forging of stronger associational linkages with the state and with labor. All business organizations want to maintain traditional business associations, in the interest of a unified front and a stronger negotiating position. It is now in the interest of big business to "renegotiate" traditional corporatist relationships rather than to dismantle them. As the evidence above suggests, by redefining, rather than dismantling, traditional relationships, big business continues to dominate economic policy-making.

Other, more obvious, attempts to undermine any movement toward real political liberalization seem to come from the PRI leadership itself. While the Zedillo regime has paid a great deal of lip service to "opening" up the political process, the PRI leadership has not implemented measures to allow for open candidate selection. The nomination process remains ambiguous and the official party is split.

Most indicators seem to point to a struggling democracy, but one that feels more comfortable with some elements of corporatism, again, a *redefined*, "new" version. Consistent with traditional political strategy, mechanisms to assure the succession of the PRI have not yet been completely dismantled. A series of political assassinations, kidnapings of wealthy businessmen, increasing social activism in the South, the marginalization of a greater percentage of the population, and the growth of absolute poverty make the prospect of democracy a distant one.

Notes

1 Carlos Arriola Woog, *Los Empresarios Y el Estado, 1970–1982* (Mexico: Miguel Angel Porrúa Librero Editor, 1988), p. 38. For reference to V. Carranza see Robert E. Scott, *Mexican Government in Transition*, rev. edn (Chicago, IL: University of Illinois Press, 1971), p. 118.
2 Ibid., pp. 38–9.
3 See Judith Hellman, *Mexico in Crisis* (New York: Holmes & Meier, 1987), p. 15.
4 See G. Roberto Newell and F. Luis Rubio, *Mexico's Dilemma: The Political Origins of Economic Crisis* (Boulder, CO: Westview Press, 1984).
5 See Charles C. Cumberland, *Mexico: The Struggle for Modernity* (London: Oxford University Press, 1976), pp. 269–72.
6 For a more detailed analysis of the student movement see Hellman, op. cit., pp. 131–5.
7 For a complete discussion see Remonda Bensabat Kleinberg, *Strategic Alliances and Other Deals: State–Business Relations and Economic Reform in Mexico* (Durham, NC: Carolina Academic Press, 1999), Ch. 4.
8 Rogelio Hernandez Rodriguez, "Empresas y Empresarios en la Sexenio de Miguel de la Madrid," in Celso Garrido N. (ed.), *Empresarios y Estados En America Latina* (CIDE, Fundacion Freidrich Ebert, Mexico, 1988), pp. 71–3.

9 Sylvia Maxfield, "The International Political Economy of Bank Nationalization: Mexico in Comparative Perspective," *Latin American Research Review*, **27**, 1 (1992), 98.
10 Ibid., p. 72.
11 Ibid., pp. 78–9.
12 Grupo Desc, Sociedad De Fomento Industrial, SA de CV, y Companias Subsidiarias, *Annual Report*, 31 December 1991, Mexico, p. 4.
13 *Expansion*, 21 August 1991, p. 312.
14 Ibid., pp. 368–9.
15 *Business Mexico*, **11**, 3 (March 1992), 16.
16 *El Financiero*, 2 April 1992.
17 *The Mexico City News*, 22 May 1992.
18 *El Financiero*, 15 April 1992.
19 *Business Mexico*, **11**, 3 (March 1992), 16–7.
20 *The Mexico City News*, 22 May 1992.
21 *Business Mexico*, **11**, 3 (March 1992), 16.
22 *Unomasuno*, 11 May 1992.
23 *Business Week*, 22 July 1991, pp. 40–2; see also Judith Teichman, *Privatization and Political Transition in Mexico* (Pittsburgh: University of Pittsburgh Press, 1996), p. 265.
24 *Unomasuno*, 23 May 1992.
25 By July of that same year there were 15 total financial grupos in operation, *La Jornada*, 20 April 1992, p. 22.
26 *El Financiero*, 4 May 1992.
27 *El Universal*, 7 July 1992, Section F, 1.
28 After receiving authorization to operate as a financial group on 28 May 1991, and acquiring controlling shares of Bancomer, Grupo Financiero Monterrey changed its name to Grupo Financiero Bancomer, *La Jornada*, 7 July 1992, 38.
29 *La Jornada*, 7 July 1992, 38.
30 Interview, Lic. Alejandro Paez Aguilar, Director of Promotion and Public Relations for Consultores Internacionales, Mexico City, 26 June 1992.
31 Ibid.
32 See Kleinberg, *Strategic Alliances and Other Deals*, Ch. 6.
33 Ibid.
34 Interview, Lic. Armando Ruiz Galindo Urquidi, Vice President, ANIERM; Director of International Affairs, CANACINTRA, Mexico City, 13 July 1992.
35 Interview, Oscar Vera, Economist, Centro de Estudios Economicos del Sector Privato (CEESP), Mexico City, 12 May 1992.
36 Interview, Javier Ortiz de Montellano, Senior Economic Advisor for the Canadian Embassy, Mexico City, 29 May 1992.
37 *El Financiero*, 7 May 1992.
38 *El Financiero*, 13 May 1992.
39 *El Financiero*, 7 May 1992.
40 *El Financiero*, 14 April 1992.
41 *La Jornada*, 12 May 1992.
42 *El Financiero*, 21 May 1992.
43 *La Jornada*, 3 July 1992.
44 *El Financiero*, 12 June 1992.

45 Interview, Jose Luis Mastretta Galvan, Director of Business Thinking, CANACO–Monterrey, Monterrey, Mexico, 18 July 1996.
46 Ibid.
47 Interview, Roman Vidal Tomayo, Foreign Trade Director for CANACO, Mexico DF, 25 June 1996.
48 Like the "pacts" before it, referred to by the de la Madrid and the Carlos Salinas governments as the *Pacto*, the *Allianza* was a way of involving both the labor sector and the private sector in inflation-fighting strategies which include wage reductions and freezes, more extensive deregulation in various sectors, etc. Typically, government, labor and business leaders meet on a bi-monthly basis to negotiate and monitor its progress.
49 Robert Kaufman, *The Politics of Debt in Argentina, Brazil, and Mexico* (Berkeley, CA: University of California Press, 1988), pp. 96–7.
50 Ibid., p. 93.
51 Mexican Statistical Agency, *INEGI*, May 1996.
52 Ibid.
53 Ibid.
54 That the private sector did not want to see the dismantling of unionism in Mexico was a sentiment expressed by all business organization presidents interviewed in June and July of 1996.
55 Vivianna Patroni, "It Takes Three to Tango: Capital, Labor, and State in the Mexican Labor Legislation Reform Dispute." Paper presented at the Annual Meeting of the Canadian Political Science Association, Learned Societies Conference, Brock University, St Catherines, Ontario, June 1996, 14.
56 Interview, Julio Bolvinik, Poverty Economist, El Colegio de Mexico, 5 July 1996.
57 Julio Moguel, "Salinas Failed War on Poverty." Website article.
58 Bolvinik, op. cit.
59 Ibid.
60 Ibid.
61 Ibid.
62 Ibid.
63 Ibid.
64 Phillippe C. Schmitter, "Transitology: the Science or the Art of Democratization?," in Joseph S. Tulchin and Bernice Romero (eds), *The Consolidation of Democracy in Latin America* (Boulder, CO: Lynne Reinner, 1995), p. 15.
65 Luis Alfonso Ramirez Carrillo, "Old Links, New Brokers: State–Society Relationships in Modern Mexico." Presented at the Society for Latin American Studies (SLAS) Annual Conference, St Andrews University, Fife, 4–6 April 1997.
66 Ibid.
67 Ibid.
68 Roderic A. Camp, "Political Liberalization: the Last Key to Economic Modernization in Mexico?," in Riordon Roett (ed.), *Political and Economic Liberalization in Mexico: At a Critical Juncture?* (Boulder, CO: Lynne Rienner, 1993), p. 18.

11
Democracy and Organized Labor in Argentina: Challenges and New Alternatives?[1]

Viviana Patroni

Introduction

Since 1983 Argentina has experienced what is by now the longest democratic period in its history, and it has done so while at the same time facing a series of unparalleled economic crises involving recession, hyperinflation, de-industrialization and, from the end of the 1980s, a radical process of restructuring. As in the rest of Latin America, restructuring in Argentina has meant the implementation of a painful and not always coherent program designed to alter the basis upon which the economy had been previously organized. In what might represent yet another in the long list of paradoxes that characterized the country, restructuring in Argentina was the result of the coming to power of a government that not only campaigned on a platform based on appeals to traditional distributional concerns, but that had also been the political force that crystallized the previous development pattern.

The reforms introduced by the current Peronist administration of President Menem, now in his second term, thus run contrary to the logic of the alternative that until the 1970s propelled the country's economy. Inward-orientated industrialization, a key role for the state in the regulation of prices and salaries, and a large public sector were erected by Perón in the 1940s as the pillars of an economic alternative that also changed the political map of the country in a profound fashion. In fact, the emergence of Peronism into Argentine politics is one and the same process as the irruption of labor as a major political actor in the country. So powerful was the impact of one on the other that, at least until recently, it would have been as difficult to imagine a non-Peronist labor movement as a Peronist party divorced from the discourse of workerism.

The transformation that Argentina has experienced under Menem confronts the labor movement with two major conflicts. One is related to the economic impact of restructuring on workers, particularly declining wages, de-industrialization, structural heterogeneity, and unemployment. Besides their effects on welfare, these outcomes of restructuring have had a major and negative bearing on the actual power of labor organizations. The other is the result of the transformation of Peronism implicit in its determination to implement such changes. To a large extent Menem's success in introducing his economic reform program can be accounted for by his political credentials. Being the president from the party with a strong historical claim as the main political referent of the working class has allowed Menem to surmount one of the most tenacious opponents that all regimes since the 1950s encountered: the Peronist labor movement.

In this chapter I am going to focus precisely on the experience of this key and undeniably still majority sector of the organized working class. My argument will be first that organized labor's structural weakness and its shared political identity with the government have not only affected its response to neoliberal reforms but have also, and much more importantly, dramatically altered the basis upon which its power as a corporatist actor rested. However, this has not implied the total disappearance of the power of the Peronist union structure or of the appeal of corporatism as an effective form of controlling labor's oppositional power.

Further, I will argue that although democratization in the 1980s involved an important questioning of the authoritarian nature of key corporatist actors in Argentina, the material requirements of structural reforms have limited the political appeal of a frontal attack on labor bureaucracies. Moreover, while structural reforms have weakened the power of traditional labor leaderships they have also made it difficult for other sectors within the labor movement to effectively challenge the exclusionary nature of that sector's leadership.

In order to elucidate these points the discussion will be first directed at tracing the emergence of Peronism as the hegemonic force within the labor movement and its significance in consolidating the power of organized labor as a corporatist actor. Then, in the second part, I will deal with the aftermath of Peronism and the transformation of the labor movement into a major political force in Argentina. This background will permit a consideration of the extent to which democracy and restructuring have posed themselves as challenges, and also as new opportunities for the still powerful Peronist labor movement.

Perón and the Argentine labor movement

The emergence of Peronism as the dominant political identity of the working class represents a watershed in the history of the Argentine labor movement. The most immediate antecedent to the consolidation of Peronism in Argentina can be found in the early and rapid process of industrialization that took place during the 1930s. Particularly after the middle of this decade, industrial growth accelerated under a project of limited industrialization that, contrary to what became the norm in the rest of Latin America, was supported by the traditional elites.[2] This process of industrialization is central in all accounts of the rise of Peronism in Argentina

By the mid-1940s the number of workers in industrial plants had doubled and, for the first time, the manufacturing sector contributed a larger share of GDP than agriculture. More importantly, industrialization during this stage excluded from its main goals any form of income redistribution. This pattern of economic exclusion, added to its political marginality, created a unifying experience for the working class and a common identity that transcended its internal divisions.[3] Part of the common experience of this working class, or more precisely its organizations, had been the rather ineffectual results of their struggles. This was what Perón could offer to change.[4]

Because Peronism is such a complex issue one always risks a degree of over-simplification in accounting for its emergence and presence in Argentine politics.[5] Nevertheless, let us attempt to reach a tentative summary regarding what Peronism meant for labor. Several variables intervene in the political equation that made possible the consolidation of Peronism. The power of the state to deliver some of labor's long-standing demands, to articulate a project of development involving an improvement in the distribution of income, to implement legislation guaranteeing fundamental labor rights and to promote a momentous social reform, were key in gaining the support of labor. As happened in many other countries, one central piece of legislation involved transferring to the state the power to recognize the legality of unions.[6] When these opportunities were not enticing enough on their own, Perón did not hesitate to use various forms of coercion to bend the will of more recalcitrant unions.[7]

It is also important to consider what Peronism did not imply. The first characteristic worth noting is that labor was not some form of "natural" ally in the rise of Perón to power. On the contrary, the relationship developed on the basis of mutual distrust and after a process of complex maneuvering. Moreover, Perón was also cautious to find counterbalances to the participation of labor into the alliance he was trying to build. The rise of Perón to power, and the role that labor unions played in it, also serve to illustrate the struggle that they presented to retain their independence and autonomy. Thus, their eventual absorption into the movement that demanded their total compliance to Perón can be seen as a kind of defeat at least to the extent that it represented the outcome of a political situation that the leadership of the labor movement attempted consciously to prevent.

The position of labor became increasingly subordinated to Perón through his two presidential terms between the electoral victory of 1946 until he was overthrown in a military coup in 1955. During his two presidencies Perón introduced important new labor and social legislation that translated into major gains for the working class. He granted constitutional recognition to unions, but did not guarantee the right to strike. Wages also increased substantially in relationship to other sectors, from 38.3 percent in 1935–36 to 46.4 percent by 1953–55.[8] Organized labor grew enormously during the period. From approximately half a million members in 1945 it reached 2.5 million in 1954, an almost 5-fold increase during a decade. Those unions supported by the government benefited not only in terms of their size but also of their power. Moreover, once union dues became compulsory, unions became financially powerful, able to provide handsome salaries and other benefits for their leaders. Not only did the material rewards of leadership increase, but there was a new and important sense of prestige attached to such positions, their power usually involving access to political appointments and elected offices.[9]

The power that labor acquired during this period allowed it to become one of the central players in the unfolding of the political conflicts that would characterize Argentina after the downfall of Perón. To the structural advantage of low rates of unemployment, Peronism added to the organizations of the working class two critically important features for their success as political actors: political coherence and a strong organization.[10]

State and labor after Perón

The military coup of 1955 ("The Liberating Revolution") represented an important point of transformation in Argentine politics. Peronism had given the working class a new political identity that resulted in a deep polarization of society and thus set the stage for the cycles of economic and political crisis that with increasing violence dominated the country for the following three decades. The conflict centered on the direction that industrialization and state involvement should take, but neither of the two segments that emerged in the aftermath of Peronism – the populist alliance that Perón himself had brought into existence and the modernizing, "developmentalist" alternative that opposed it – ever questioned these two basic premises in the development pattern followed by Argentina since the 1930s.[11] Their overall agreement regarding the necessity of industrialization and of state intervention, however, generated only a very tenuous consensus that was overcome periodically by the distributional struggles of labor and capital.

The conflict over the distribution of the gains from development never involved, though, an actual challenge to the power of the Peronist labor structure. On the contrary, even the anti-populist coalition supported the expansion of unions' organizational and financial capabilities.[12] Thus, the state protected labor's corporatist structures by ratifying legislation that was at the core of its power: the principle of a single union per trade and the monopoly of representation at the national level for the General Confederation of Labor (CGT). Since 1969 the prerogatives of the recognized union in a branch of activity also involved the institutionalization of its power to control and administer health-care services for its members.[13] But the enormous capacities acquired then by the organizations of labor were in direct relation to the power that the state held over them. It was the state that could grant, suspend, and cancel the status of a union as the only organization with the right to negotiate collective agreements *(personería gremial)*. Thus it was also the state that determined the legal right to have access to the funds administered through welfare programs (*obras sociales*) since only the recognized union could provide these services. In other words, the state was central in the maintenance and expansion of the corporatist power of unions and thus most of their struggles were directed at it.

Becoming a "political factor" was a function of the concessions that the labor movement could extract from the state and, through it, from capital. But the labor movement did not enter the game empty-handed. Peronism had created coherence within the movement and legitimacy for its leaders. The incorporation of labor unions into the political coalition that brought Perón to power, and the acceptance of the legal framework regulating unions' activity he imposed, had set the limits to what constituted politically feasible actions. However, both also represented the channels through which labor had gained access to the state, and this was the key side of the equation. Labor did occupy institutional and political spaces that altered in an unequivocal fashion the relationship among labor, capital and the state.[14]

The period between 1955 and 1976 can then be understood for its centrality in the re-configuration of a labor movement with an enormous power as a corporate actor. It was not strange then that, during this time, the political lifespan of military and civilian governments was to a large extent determined by organized labor. When labor leaders openly confronted the goals of civilian administrations the conflict usually led to a new military intervention that displaced the elected government. But the room for military regimes to defy labor opposition was equally limited. Whenever this reached levels only controllable by very high levels of repression, at least until 1966, military rulers opted to withdraw from power and call new elections.[15]

Neither the three military regimes (1955–58, 1962–64, 1966–73) of the period, nor the civilian governments (1958–62, 1964–66, 1973–76) were successful in finding a resolution to the impasse that permeated Argentine politics. However, as crises worsened, Peronism itself became a much more heterogeneous force which eventually translated into growing divisions within labor. The conflict among these factions acquired violent characteristics that reached their most chaotic expression after the death of Perón in 1974, while he was once again the President of Argentina. With the return of Peronism to power, the CGT – that also represented the most conservative and orthodox bloc within the party – used all the resources at hand in order to eliminate the most radical and oppositional sectors within and outside the Peronist Party. Its power was also central in obtaining congressional approval for a new law of professional association (*Ley 20.615*) that became a key instrument in displacing dissenting fractions within unions as it strengthened the centralized and authoritarian organization of unions and the CGT.

In a climate of economic and political chaos the military took power once again in March of 1976. Like no other military dictatorship before, it was determined to uproot once and for all what it perceived to be the causes of Argentina's problems. According to its diagnosis the social catastrophe that afflicted Argentina was the result of the pattern of development followed until then that had increased the power of labor and sparked the rise of "subversion" in the country. According to this premise the primary targets of repression were left-wing unionists from the dissident sectors that had challenged the role of traditional union bosses. For them the state reserved its most brutal expressions of violence: death and disappearance. But even orthodox union leaders suffered at the hands of the new dictatorship that did not hesitate to employ a heavy hand against the CGT. Not only were some of its most prominent leaders jailed, but the organization itself was dismantled and unions' social welfare programs transferred to the state.

As economic plans faltered in the early 1980s, and its political allies withered, the military junta attempted to recoup support by embarking the country on a nationalist war against Great Britain. It was a desperate venture that led, with hindsight, to a not-so-surprising defeat. When this event was added to the long list of grievances Argentina held against the dictatorship, it collapsed precipitously, forcing the military to call for elections in 1983.

The return to democracy: Alfonsín and the labor movement

The return of democratic rule in Argentina confronted the Peronist labor leadership with the necessity of finding accommodation to radically different conditions. To begin with the economic situation of the country had changed dramatically during the years of dictatorship. One of the most adverse outcomes of the military's neoliberal policies, specifically the combination of trade liberalization and overvaluation, was the significant de-industrialization of the country. Industry had represented 28.2 percent of GDP in 1974, but its share had fallen to 25.3 percent in 1978 and 22.4 percent in 1982.[16] One consequence of this process of de-industrialization was the growth of unemployment and underemployment. The distribution of income had also suffered a regression and real wages had fallen consistently, particularly during the first years of dictatorship. If these structural changes had created new challenges for the union movement, the restrictions placed by the military on union activity in general, and on the Peronist labor leadership in particular, had profoundly hindered the labor movement's ability to resist the regime.[17] Moreover, the military had once again produced major divisions that also contributed to its weakness.[18]

But the political context within which democratization was to take place was no less of a challenge for Peronist unionists: for the first time in its existence the Peronist party – the hegemonic political force within the CGT – was to lose a free and unrestricted election in 1983. The electoral defeat of Peronism carried with it major implications for Peronist labor leaders since they had been central figures in the reorganization of the party. In addition, the basic working premise of Raúl

Alfonsín – the newly elected president – was that he could trigger a process of renewal in the labor movement, weakening the traditional labor orthodoxy, and thus increasing his chances of addressing successfully the problem that all regimes in Argentina had confronted since 1955: how to neutralize the power of unions to destabilize the government.[19] Accordingly, Alfonsín's first act in power was to submit to Congress a draft bill containing a proposal for the democratization of unions *(Ley de Reordenamiento Sindical)*.[20]

This attempt on the part of the Executive prompted the rapid and negative response of the CGT's leadership, that quite accurately evaluated the maneuver as an open challenge to its power. Although the labor movement had been divided during the dictatorship, one of its first accomplishments in the early post-military period was to restore its institutional unity. The task was somehow simplified by the fact that the repression unleashed under the military against the most progressive sectors of labor had freed the Peronist union structure of one of its major problems. Its reunification was an effective and unambiguous signal to the new democratic government that the CGT was well on its way to recovering its traditional potential for opposition.[21] Whatever the limitations of this labor movement, in terms of its bureaucratization, authoritarianism, lack of prestige, and corruption, it was still a potent actor not to be dismissed.

The Executive's draft bill for a new law of professional associations was approved in the lower chamber of Congress but defeated by the Senate where the Radical party did not have a majority. This important defeat for Alfonsín set the conflict with the Peronist labor movement in a new framework. Given that negotiations with labor were inescapable, the government retreated then to the traditional pattern of concessions to the leadership, protection of its corporatist power, and, of course, the abandonment of those initiatives that could worsen the already-tense relationship.[22]

Consequently, after this first major setback, Alfonsín had to accept the renegotiation of the new law of professional associations with the CGT. This law was approved only in January 1988, but it secured the continuation of the legal provisions that had protected the corporatist power of the peak labor organization. It also eliminated the restrictions on the right to strike, particularly to the so-called "political strikes," that the Radical administration had sought.[23] The defeat of the bill also implied a retreat from the plans Alfonsín had harbored regarding changes in the administration of health and welfare services. After a long process of negotiation, unions were allowed to regain control of this key component of their financial might, agreeing in exchange to accept public reviews of their activities.[24]

The conflict between Alfonsín and labor was framed within the context of the unprecedented economic crisis inherited from the previous military regime that conditioned the transition to democracy in a critical fashion. Insofar as the economic policies applied by Alfonsín did not produce the expected results, the need to gain some time and political space became crucial for the new administration. In the hope of securing those essential conditions for their survival, and of compromising other social actors with the success of transitional policies, Alfonsín included them in a process of tripartite negotiations or *concertación*.[25]

Until 1985 the issue of social and economic *concertación* was vital to government policy. Nevertheless, by June of that year the government abandoned the primacy given to negotiation and, instead, took an independent stand by proposing a new plan of stabilization, the "Plan for Economic Reform," better known as "*Plan Austral*." The main objective of the plan, as was common in "heterodox" approaches, was to control inflation without producing a recession, but the *Plan Austral* also marked an important shift in the approach that the government had followed with respect to corporatist actors. Up to this point Alfonsín had responded to their demands to be included in a process of *concertación*, but the new plan was clearly the exclusive initiative of the state.[26]

The initial success of the plan in reducing inflation created the much-needed space that the government had been seeking. Moreover, the new electoral victory in the Parliamentary elections of November 1985 helped it retain the economic and political impulse required to neutralize any threat to the survival of democracy. However, the relief was only temporary. In response to the new economic program, and also as part of its tactics as a major political force in the country, the CGT stepped up its opposition to the government by walking away repeatedly from the negotiating table and by staging an increasing number of general strikes during 1986.[27] The incompatibility between the political objectives of the Radical administration and the Peronist labor leadership became painfully evident for Alfonsín as subsequent stabilization plans stumbled, particularly at the moment when wage negotiations entered into the equation.

The CGT maintained an openly confrontational position with respect to the government, insisting on wage increases and employment stability. When the spaces opened by the government for negotiation were not sufficient to display its corporatist strength, the CGT did not hesitate to make use of general strikes, one of its classical instruments for pressuring the government. During the administration of president Alfonsín the CGT called 13 general strikes. Moreover, the confederation went as far as finding common ground with the organizations of capital in order to gain greater leverage.[28]

Nevertheless, a constant obstacle in the effectiveness of labor's actions was the internal divisions in the movement. Some of the major factions within the CGT during this time echoed quite closely the long-standing divisions prevalent within the labor movement since the 1960s. However, the transition to democracy also witnessed the emergence of some important realignments within the Peronist labor movement. Most notably there were some important new alliances developed among sectors that had previously existed as opposing forces. Regardless of these divisions, there was no question that Peronism continued to be the hegemonic force within the labor movement.[29]

In a general sense the CGT's political behavior in the period of transition responded to its hesitations regarding the benefits it could derive from facilitating the political success of the Radical party. On the contrary, for the CGT it was much more important to guarantee the victory of the Peronist party in the future election, a victory that labor bosses equated with a greater access to the spheres of power within the state. This became an even more important variable after the electoral success of the Peronist party in the congressional elections of September 1987. Nevertheless, opposition of the government was expressed on several occasion in terms that raised questions about organized labor's commitment to democracy. The repeated references to the classic theme within the Peronist union leadership regarding the inadequacy of democratic forms that did not respond to the priorities of "social justice" created concerns in moments when democracy seemed fragile at best.

It is important to mention, though, that Alfonsín's original impetus in counteracting the authoritarian nature of labor organizations and allowing more democratic forms of rank-and-file participation in unions was also marked by ambivalence. The government needed to count on well-structured social actors capable of controlling the response of their membership to the agreements that Alfonsín had hoped to negotiate. The alternative of a transformation in the organization of unions and the new internal competition implied in a renewal of traditional leaderships could pose a further threat to the process of *concertación,* one that Alfonsín eventually feared and clearly tried to avoid.[30]

The reorganization of Peronism

The political calculations of the CGT regarding what was at stake in its relationship with the Radical administration cannot be seen in isolation from the role of its different factions in the recomposition of the Peronist party (Partido Justicialista [PJ]). Its defeat in the elections of October 1983 raised some serious questions about the role of organized labor in the determination of the party's tactics. Because the "orthodox" faction of labor was at that time the hegemonic force within the party, the conflict emerged precisely over its capacity to reconstitute itself as a winning force in the future. The questioning also involved an evaluation of the basis upon which Peronism could democratize its internal structure and limit the role that unions had played in its configuration. In the context of these dilemmas the party split into two main factions. One continued to be controlled by the orthodoxy from the Peronist unions, the "62 Organizaciones." But the position of political supremacy of this sector of the movement was effectively challenged after 1985 by a "renewal" faction mostly composed of "politicians" (as opposed to unionists) who counted on the support of influential union organizations, particularly the "Group of 25." Carlos Menem was one of the central figures of this sector. The renewal faction would also experience more divisions, that once again pulled factions within the Peronist labor movement in different directions, as two of its principal figures were contenders in the primary elections for the party's presidential candidate.

Probably the variable that contributed most to the success of the renewal faction was its ability to provide a means to institutionalize the broad Peronist movement into a more structured party, capable of reflecting internally a new consensus for democratic forms of representation. But success also confronted the new leadership of the party with the problem of molding the institutional change to the political identity that Peronism had previously given to all the sectors within the movement. In other words the new party consolidated since 1985 with the ascendency of the renewal faction to power needed to reconnect itself to the ideological precepts of Peronism. It was Menem who emerged as the person capable of recapturing the discursive elements traditional to Peronism. Unions figured prominently among the sectors directly addressed by his return to a traditional corporatist discourse.

The sectors within the Peronist unions that first gave their support to Menem were the representatives of unions that, while opposed to the "renewal" sectors in the party, were also inclined to a new understanding of the relationship between capital and labor and were then critical of traditional corporatist strategies.[31]

With Menem as the presidential candidate in the upcoming elections of May 1989, Peronism would make a historical comeback: but although this victory did not surprise many in the country, the new president's first acts of government and, since then, his unrelenting push for radical structural reforms was certainly baffling for many Argentines.

Structural reforms and labor

Peronism's return to power and the drastic and unexpected measures taken by the new government must be set in the context of the crisis faced by Argentina at the time of the transfer of power from the Radical administration. Alfonsín's last year as president had produced several dramatic events that would have a direct bearing on the actions of the incoming president. The military had remained a constant problem for the transitional government, but the last military uprising in 1988 and the January 1989 left-wing attack on La Tablada military base had created a new climate of political tension through which the military secured some important concessions.[32] Both events helped to instill a climate of uncertainty where what seemed at stake was the fortunes not just of the Radical government but rather of democracy itself.

The situation was not much better for Alfonsín on the economic front. In a last attempt to control inflation, Alfonsín's economic team set in motion a new stabilization program, the "Plan Primavera." However, the plan had collapsed by the beginning of 1989, unleashing at this time the final and most monumental economic crisis that any government in Argentina had ever faced. After February inflation grew at extremely high rates, particularly during the second trimester, reaching 4 923.3 percent for the year as a whole.[33] The beginning of 1989 thus witnessed an even more acute political and economic crisis, characterized by hyperinflation, drastic declines in real wages, employment, investment, and the rate of growth. Unable to exercise any degree of control over the evolution of the economy, and facing a violent wave of social mobilization, Alfonsín decided to relinquish power almost six months before his mandate officially expired.

The gravity of the circumstances under which Menem took office indisputably limited the options at the disposal of the new administration. Paradoxically, it was also the scope of the crisis that provided the state with the political ammunition it required to undertake the drastic shift in policy orientation that Menem offered as an alternative.[34]

Menem's original economic plan rested on the introduction of significant structural reforms as part of the stabilization plan. As with Alfonsín before him, at first he was only temporarily successful in controlling inflation, actually facing a hyperinflationary process again in early 1990. It was only in 1991, with the introduction of the "Convertibility Law" designed by his Minister of Finance, Domingo Cavallo, that inflation was curbed.[35] However, contrary to Alfonsín, Menem had the political power necessary to introduce key restructuring measures.

Central in the area of structural reforms was privatization, becoming in fact the spearhead for the process of restructuring as a whole. Through the enactment of the Economic Emergency Law, and particularly the Public Sector Reform Law, Menem received the constitutional powers to speed the sale of public assets.[36] Two characteristics of the process of privatization in Argentina are significant for their long-term repercussions. One is that a large portion of the capital for privatization did not originate in Argentina and, therefore, at least part of the profits will be remitted abroad.[37] Considering that many of the sectors in which this capital was invested do not produce tradeable commodities, it follows that the hard currency repatriated as profits would not have been earned through exports. The other factor pertains to the inadequate process through which the objectives of privatization were designed. The operation of privatized companies had not been properly regulated, original plans for workers' ownership in privatized companies were neglected and, most importantly, there was no attempt to establish goals to guarantee competition in the privatized sectors.

Another key area of reform was trade liberalization. However, trade liberalization has coincided with the use of an overvalued currency as an anti-inflationary instrument. Appreciation of the domestic currency has not only served as an "anchor" in the stabilization package, reducing expectations about price increases. It has also provided an instrument for relieving the heavy burden imposed on the state and the private sector as a result of their enormous level of foreign indebtedness. The combination of currency appreciation and trade liberalization has led to a very rapid growth of imports. Moreover, the overvaluation of the domestic currency has also tended to undermine the international competitiveness of exports. The overall effects of these two trends have been very negative for the industrial sector of the country. As Damill, Fanelli and Frenkel note, the process of de-industrialization and import de-substitution are responses with long-term effects to only transitory conditions of relative prices and inflow of capital.[38]

The elimination of the regulations affecting the markets for factors of production has been a central objective of the process of restructuring. In a fashion quite similar to both the military experiment initiated in 1976 and to the changes implemented since 1987, Menem moved speedily to deregulate the financial markets. At the end of the Radical administration controls over interest rate were removed. Menem also abolished a large part of the restrictions regulating the operation of capital markets. The Peronist administration also deregulated markets for goods and services, and has opened areas pertaining to social security, for instance pension plans, to private competition. Still, the deregulation of the labor markets is probably the most politically and economically provocative transformation carried forward by the Menem administration.

Reforms to the labor laws of the country were already discussed as part of the Peronist mandate in office since the electoral campaign. Obviously, and except for those sectors more closely connected to Menem, unions considered such proposals unacceptable. Later during Menem's presidency, labor reforms would take a central place within the policy package through which the government intended to find a solution to Argentina's long decline. But labor reforms were also instrumental in allowing Menem to counteract unions' responses, curtailing their power, autonomy and capacity to react against his government.

In fact, by 1996 the Peronist administration had effectively introduced several laws and executive decrees that undermined the extensive legislation that until then protected workers' rights. Some of the most serious losses involved limitations to the right to strike, reductions in public sector salaries, the legalization of several forms of precarious employment, the determination of salary increases on the basis of increments in productivity, and the drastic reduction of liabilities for work-related accidents. Menem has even challenged the monopoly of representation so far enjoyed by the CGT.[39]

It is important to note, though, that the actual implementation of these reforms did not always match the plans of the government. On several occasions legislation was blocked in congress, regardless of the Peronist majority, or was modified through negotiations with union leaders.[40]

At first the challenge of Menem was sufficient to motivate the reunification of the CGT, divided once again over its responses to the structural reforms introduced by the Peronist government. In effect the first reactions to Menem's economic policies and his maneuvering during the federation's 1989 election process, led to the division of the CGT into two competing organizations: the CGT–Azopardo, that remained loyal to Ubaldini, the official secretary of the organization, and the CGT–San Martín very closely connected to the government.[41] The separation created a CGT divided for the first time under a Peronist administration.[42] Eventually the CGT–Azopardo would suffer an internal division as the tendencies it encompassed were unable to find enough common ground, besides their opposition to Menem, to remain in a united front.

It is at this moment that the proposed new labor legislation prompted a realignment of forces in the labor movement. The reunification of the CGT in 1992, however, did not fully reconstitute the confederation. Those sectors left outside, particularly the two important unions of teachers and state employees, later organized themselves in the new Congress of Argentinean Workers (CTA) that emerged as an independent and combative alternative to oppose neoliberal reforms. The CTA has experimented with some innovative organizational forms, for instance direct membership, in order to include the unemployed and the self-employed; that is, those marginalized by the current model of accumulation. Later, in 1994, the CGT divided again as a group of more radical opponents to the policies of Menem, headed by Ubaldini, organized themselves into the Movement of Argentine Workers (MTA). This group's relationship with the CGT has been tempestuous, remaining always as a conditional partner.[43]

Divisions within the Peronist labor movement had been in part the result of deliberate government interventions in order to reduce the possibilities of effective action against it. Particularly, the government's tactic to negotiate with individual unions, as opposed to the CGT, has been an extremely useful instrument in the hand of Menem in order to create further differentiation among sectors. The government has also been ready to offer rewards for those labor organizations willing to support policy initiatives with negative consequence for other unions.[44] But, equally importantly, internal divisions have also been the outcome of the impossibility of recovering the initiative regarding a viable response to Menem and his neoliberal reforms.

The decline in unions' power is not only the result of the confusion introduced by Menem's about-face and his unexpected policies. The economic conditions of the country have changed profoundly, and they have done so in ways that are not propitious for union activity. One of the characteristics of the process of restructuring that we noticed previously was de-industrialization. Unemployment has been the most important consequence of this process. Thus, even during a short period in which the economic activity accelerated between 1991 and mid-1994, unemployment grew from 6.5 to 12 percent. But in 1995, already into a recession, unemployment reached 18.6 percent.[45] The economy recovered in 1996, but unemployment remained very high, at 17.3 percent. The combined impact of unemployment and underemployment is calculated to affect 30.9 percent of the economically active population. Because not all sectors have been affected equally, restructuring has tended to accentuate even more the tendency to heterogeneity in labor markets, again a trend present since the 1970s. To compound matters even further, new regulations such as the connection between salary increases and productivity have contributed even more to the differentiation of the working class.

Since the second half of the 1970s the labor movement has had to frame its demands within the context of decreasing wages, growing unemployment and the precariousness of labor conditions. But under Menem there are two intervening variables that have greatly increased the difficulties in organizing labor's political tactics. First, not only was the government "theirs" but also it has proceeded with reforms while retaining popular support in the polls, including Menem's reelection in 1995. Second, regardless of Menem's "transgressions" of Peronist principles, he was successful in retaining support within the party. Thus, for the Peronist labor movement the dilemma is centered around issues that are not new, but for which it had previously sought solutions through its traditional and unifying political alternative, that is Peronism. At a general level, with the closing of this alternative, the labor movement has also lost its major point of political reference and thus its capacity to mobilize its basis around an alternative project. At a more concrete level it has also confronted diminished access to its preferred interlocutor: the state.

It is important to underline that the relationship between state and labor unions has been changed, but not all the traditional links that connected the two under corporatism have been severed. Unions still retain a considerable amount of power, and while their relationship with Menem has been and continues to be conflictive, negotiation and accommodation in exchange for political support and prerogatives remain central.[46] While Menem has been able to obtain political support from sectors not traditionally allied to Peronism, especially from the employers' associations, he cannot dismiss the role of those forces that have been historically the foundation of the movement.

This has been made evident in the attempt to find new accommodation with the CGT after the mid-term electoral defeat in October 1997 and under the pressure of the upcoming presidential elections in 1999. Notwithstanding the demands of the private sector and the IMF, the government achieved the approval of a new labor law in September 1998 prepared through extensive consultations with the CGT. The new law did away with employers' expectations regarding further reductions in severance payments and the support to collective agreements at the plant level.[47] Although important labor organizations remained opposed to the new legislation, the CGT accomplished two important victories: the government's promise to protect union's health-care programs from private competition and the endorsement of industry-wide collective agreements.[48]

Conclusion: a greater scope for democracy?

The economic and political changes initiated since the early 1980s had confronted the Peronist labor movement with challenges for which it has not always been easy to find alternatives. During the early stages of democratic transition Peronist labor leaders had to accommodate themselves to a government that had uncontested popular support but that did not include them. The decision to step up the confrontation against the government of Alfonsín tended to undermine the democratic credentials that the movement could rescue from its opposition to the previous military regime. Moreover, labor's tendency to downplay the importance of democratic forms, insisting on the contrary that "democracy without social justice" was only a façade for domination, isolated labor not only from society at large, but also increasingly from key new sectors within Peronism itself.

The victory of Peronism in 1989 had been envisaged by the labor movement, until then the "backbone" of the most powerful political force in Argentina, as a new opportunity to regain the spaces it had seen evaporate under democracy. However, Menem's presidency represented one more blow to the aspirations of labor bosses. Not only were economic reforms negatively related to unions' capacity to recoup their forces, but the President also appeared decisive in his effort to undermine labor's privileged position in the party and the government. Menem did not hesitate to classify labor demands as "sectoral" while at the same time building solid bridges with the organizations of the private sector.

In what is probably the most audacious action of the government against labor unions, the current Peronist administration has introduced very radical labor reforms that have eliminated several of the key conquests labor obtained from its association with Perón. But here lies the crux of the problem behind the relationship between labor and state in Argentina today. Because the labor movement has given solid proofs of its oppositional capability, the government has seen it necessary to curb the power of unions to become the catalyst of an alternative political and economic program. But at the same time, Menem cannot afford to neglect the power that the Peronist labor movement still holds.

Thus, the relationship between Menem and the Peronist labor movement has been characterized by a two-front approach. On the one hand the government has used all its power to fracture the unity of the movement and to increase the structural diversity that might reduce its capacity to generate unified sectoral responses. On the other hand, it has been ready to reconsider those changes that – not bearing any direct benefit for the economic program – could do away with the corporatist power left in the unions. In short, the political and economic uncertainties of the process are too large to risk the support of labor for his government beyond what is strictly necessary. In the last instance the power of unions to preserve their authoritarian and hierarchical organization has also been instrumental in suffocating the emergence of more democratic and representative forms of unionism that could really pose a threat to economic restructuring.

The maintenance of some key corporatist capabilities in the hands of Peronism's long-term allies in the labor movement can prove invaluable as a form of control of more autonomous responses from the working class; but the logic does not have unlimited potential. Like the power of all corporatist labor structures, Peronist unions have been effective in maintaining their supremacy within the labor movement because they were capable of delivering concrete benefits to their members. When they could not, they took the place of an active and militant opposition. The future of the CGT as the main representative of labor interests in the country, and also as a key force within the Peronist party and government, might depend on its capacity to take a more independent stand from Menem's government. Until now, though, Peronist labor bosses have taken only timid steps in this direction.

Thus we can conclude by underlining the rather problematic and ambiguous nature of the relationship between labor and democracy in Argentina. The uneasiness in the relationship has at its core a two-faceted conflict. One is rooted within the movement itself, since the development of a strong bureaucracy with enormous powers over the determination of the actions of unions took place at the expense of the existence of more participatory and democratic organizational structures. This not only assumed the form of an increasing distance between leadership and rank-and-file but at some points it acquired violent characteristics, particularly as challenges to the exclusionary nature of unions mounted in the 1970s. The other pertains to the role of unions within society at large. Not only was the Peronist labor leadership not committed to internal democracy, but the defense of democracy in society became at best a secondary issue. The organizing principle in the confrontation between unions and the CGT against the state was the preservation of their power for which democracy was not always a requirement. Surely enough, organized labor's tepid commitment to democracy was paralleled by the behavior of all corporatist players in Argentina and some key political actors as well.

Ultimately, the question about the relationship between organized labor and democracy should be posed in terms of the scope and significance of democracy in Argentina today. For the original identification with the ideal of overcoming the authoritarian nature of the country's political practices gave way to the ever more pressing and urgent need to address a deepening economic crisis and the concomitant deterioration in the standard of living of a growing number of Argentines. The economic alternative embraced as the only viable response to this crisis has set in motion a deep process of economic change that so far has not given momentum to a parallel transformation of political forms. Thus, while the most severe expressions of authoritarianism appeared to have been surmounted, effective spaces for democratic participation continue to be an elusive reality, the promised land that neoliberalism might just not be able to deliver.

Notes

1 The author gratefully acknowledges that financial support for this research was received from a grant partly funded by Wilfrid Laurier University Operating funds, and partly by the SSHRC general research grant awarded to WLU.

2 For a discussion about the timing of import substitution in Argentina see Javier Villanueva, "El origen de la industrialización Argentina," *Desarrollo Económico*, **12**, 47 (October–December 1972).

3 Miguel Murmis and Juan Carlos Portantiero, *Estudios sobre los orígenes del peronismo* (Buenos Aires: Siglo Veintiuno, 1971), p. 76.
4 Regarding this point it should be clear also that Perón did not need the support of unorganized workers but rather of the leadership of the established unions, since they were the ones who could guarantee a maximum gain. Juan Carlos Torre, *La vieja guardia sindical y Perón: Sobre los orígenes del peronismo* (Buenos Aires: Editorial Sudamericana, 1990), pp. 67–7.
5 Perón's rise to power was connected to the military coup that in June 1943 brought to an end the period of conservative rule initiated in 1930. From his initially obscure position within the regime, in a very short period of time Perón would gain prominence. As Secretary of Labor Perón was to build an alliance with the organized labor movement that only three years later guaranteed his electoral victory and the cementing of his development program.
6 In October of 1945 Perón introduced the new Law of Professional Associations. In a typical corporatist fashion the state acquired the power to grant the legal right (*personería gremial*) to only one union per sector to negotiate collective agreements and to guarantee monopoly to the union thus constituted within its industry or economic sector (*sindicato único*).
7 Ruth Berins Collier and David Collier, *Shaping the Political Arena: Critical Junctures, the Labor Movement and Regime Dynamics in Latin America*, (Princeton, NY: Princeton University Press, 1991), p. 339 and Hugo del Campo, *Sindicalismo y peronismo, Los comienzos de un vínculo perdurable* (Buenos Aries: CLACSO, 1983), p. 184.
8 David Rock, *Argentina, 1516–1987: From Spanish Colonization to Alfonsín* (Berkeley, CA: University of California Press, 1987), p. 263.
9 Ibid., p. 284, and Collier and Collier, op. cit., pp. 341–2.
10 Ricardo Gaudio y Andrés Thompson, *Sindicalismo peronista/gobierno radical: Los años de Alfonsín* (Buenos Aires: Fundación Friedrich Ebert, 1990).
11 See Guillermo O'Donnell, "States and Alliances in Argentina, 1956–1976," *Journal of Development Studies,* 15, 1 (October 1978) 3–33; Marcelo Cavarozzi, "Political Cycles in Argentina since 1955," in Guillermo O'Donnell, Philippe C. Schmiter and Laurence Whitehead (eds), *Transitions from Authoritarian Rule: Latin America* (Baltimore: Johns Hopkins University Press, 1986).
12 Cavarozzi, "Political Cycles in Argentina since 1955," p. 43.
13 Omar Moreno, "La última oportunidad del sindicalismo argentino," in Holm-Detlv Köhler and Manfred Wannöffel (eds), *Modelo Neoliberal y Sindicatos en América Latina* (México: Fundación Friedrich Ebert, 1993), pp. 53–6.
14 Marcelo Cavarozzi, *Sindicatos y política en Argentina* (Buenos Aires: CEDES, 1984), pp. 16–17.
15 Cavarozzi, "Political Cycles in Argentina 1955," p. 31.
16 Marcelo Diamand and Daniel Naszewski, "Argentina's Foreign Debt: its Origin and Consequences," in Miguel S. Wionczek and Luiciano Tomassini (eds), *Politics and Economics of External Debt Crisis: The Latin American Experience* (Boulder, CO: Westview Press, 1985), p. 243. See also Bernardo Kosacoff and Daniel Azpiazú, *La industrial argentina: desarrollo y cambios estructurales* (Buenos Aires, Centro Editor de América Latina, 1989), p. 16.

17 It is important to mention, though, that the labor movement in Argentina maintained high levels of communication with the military when compared to the experience of its counterparts in Brazil, Chile or Uruguay. Paul W. Drake, *Labor Movements and Dictatorships: The Southern Cone in Comparative Perspective* (Baltimore, MD, and London: Johns Hopkins University Press, 1996), pp. 163–4.

18 Although the organized labor movement in Argentina remained united in its identification with Peronism, the tactics toward the military regime split the CGT once again.

19 Andrés Thompson, "De la primavera populist and invierno conservador: El sindicalismo peronista entre 1973 y 1989," in Moreno, op. cit., p. 87.

20 Paul G. Buchanan, *State, Labor, Capital: Democratizing Class Relations in the Southern Cone* (Pittsburgh, PA: University of Pittsburgh Press), 1995, pp. 136–7, and Thompson, op. cit., p. 87.

21 Julio Godio, *El Movimiento Obrero Argentino (1955–1990): De la resistencia a la encrucijada menemista* (Buenos Aires: Legasa, 1991), pp. 392–3.

22 Thompson, op. cit., pp. 87–8.

23 Buchanan, op. cit., p. 138.

24 Ibid., pp. 141–2.

25 Following De Riz, Cavarozzi and Feldman, by *concertación* we will understand the policy-setting mechanism established through the participation of labor and employers' associations and conducted in the context of regular, planned, and long-term consultations between these organizations and the state (De Riz, Cavarozzi and Feldman, op. cit., p. 7)

26 Ibid., p. 51.

27 Héctor Palomino, "Los sindicatos bajo el gobierno constitucional: de la confrontación a la alianza," in José Nun and Juan Carlos Portantiero (eds), *Ensayos sobre la transición democrática en la Argentina* (Buenos Aires: Puntosur, 1987), p. 183.

28 Thompson, op. cit., p. 89. The "Group of 11" (encompassed by the labor's confederation and 10 organizations of the private sector) was the formal sphere where the CGT joined the employers' association in an attempt to increase the pressure on the Alfonsín administration. See also Juan Carlos Portantiero, "La concertación que no fue: de la Ley Mucci al Plan Austral," in Nun and Portantiero (eds), op. cit., pp. 162–3.

29 Most notably, in union elections carried out between 1983 and 1984, Peronism appeared as the unquestionable winner. In fact, in 90 percent of union electoral processes Peronism was victorious (Godio, op. cit., p. 401).

30 Palomino, op. cit., p. 178.

31 The previous paragraph is drawn from the work of Vicente Palermo and Marcos Novaro, *Política y poder en el gobierno de Menem* (Buenos Aires: Norma, 1996), pp. 186–98.

32 In January 1989 a left-wing group (*Movimiento Todos por la Patria* – MTP) attacked a military garrison at La Tablada, apparently in the hope of preventing a military coup being planned by the same group that had rebelled against Alfonsín only a month before. See J. Patrice McSherry, *Incomplete Transition: Military Power and Democracy in Argentina* (New York: St Martin's Press, 1997), pp. 190–2).

33 Mario Damill and Roberto Frenkel, "Malos tiempos: la economía argentina en la década de los ochenta," in Roberto Frenkel *et al.* (eds), *Argentina: evolución macroeconómica, financiación externa y cambio político en la década de los 80* (Madrid: CEDEAL, 1992), p. 63.
34 Palermo and Novaro, op. cit., p. 115.
35 Though the "Convertibility Law" the government established a fixed parity between the dollar and the domestic currency. The new law also established that the monetary base must be guaranteed 100 percent by reserves either in foreign currencies or in gold. See Paulo Nogueira Batista Jr, "Crisis monetaria, dolarización y tipo de cambio," *Revista de la CEPAL, 1993,* no. 50, pp. 101–2.
36 Through the Economic Emergency Law the government was allowed to suspend practically all subsidies to the private sector and other government expenditures on a wide array of programs such as export subsidies and some regional transfers. The Public Sector Reform Law set up the main regulatory framework for the privatization of state-owned companies. As mentioned, one fundamental aspect of the laws was to allow the government to acquire extraordinary powers in the management of the crisis.
37 Daniel Azpiazú y Eduardo Basualdo, "Las privatizaciones en la Argentina. Evolución de los precios relativos e impacto macroeconómico" (Buenos Aires: IDEP, July 1995) (mimeo), p. 32.
38 Mario Damill, José Maria Fanelli and Roberto Frenkel, "De México a México: el desempeño de América Latina en los noventa," *Desarrollo Económico,* **36**, (Summer 1996), p. 71.
39 Moreno, "La última oportunidad del sindicalismo argentina," pp. 63–5; Manzetti, op. cit., pp. 236–7; and Ricardo Falcón, "Políticas neoliberales y respuestas sindicales (1989–1992)," in Moreno, op. cit., p. 96.
40 Sebastián Etchemendy, "¿Límites al decisionismo?," in R. Sidicaro *et al.* (eds), *Política y sociedad en los años del menemismo* (Buenos Aires: UBA, 1995).
41 Manzetti, op. cit., p. 234.
42 Falcón, op. cit., p. 97.
43 Thus, for instance, the introduction of some new decrees on the issue of labor legislation produced some new tension between the CGT and the MTA regarding the most appropriate response (*Clarín,* 6 January 1997, p. 5).
44 Jeremy Adelman, "Post-Populist Argentina," *New Left Review,* **203** (Jan/Feb 1994), pp. 83–4.
45 Damill, Fanelli and Frenkel, op. cit., p. 72. Economic growth in Argentina reached 8.9 percent in 1991, 8.7 percent in 1992, 6 percent in 1993, 6.5 percent in 1994, but fell by –4.4 percent in 1995 (*Clarín Económico,* 28 July 1996, p. 4).
46 Falcón, op. cit., pp. 102–4.
47 *Clarín,* 24 April 1998.
48 *Clarín,* 28 April 1998.

12
Neo-Liberalism, Democratization and the Women's Movement in Nicaragua

Katherine Isbester

Introduction

In 1979 a socialist revolutionary party, the Sandinista Front for National Liberation (FSLN), took power in Nicaragua. The FSLN had an ideology of social justice and mass political participation within a corporatist state. Women benefited from subsidized food, improved legal rights, free health care and education, and a government sympathetic to women's needs. The government organized women into their own organization, the Nicaraguan Association of Women Luisa Amanda Espinoza (AMNLAE). AMNLAE had a seat in the Nicaraguan National Assembly with other popular organizations. Despite the efforts of grassroots women and some of their leaders, AMNLAE remained hierarchical, non-democratic, and controlled by the FSLN. As a result women's participation in their own organization declined throughout the 1980s.

In 1990 the Nicaraguan electorate replaced the FSLN with a right-wing party led by Violeta Chamorro. The Chamorro government privatized the economy through a harsh neo-liberal program. This rapidly eroded the gains that women had achieved under the FSLN. Their standard of living fell, their employment opportunities diminished, and their access to remaining government-funded programs and services was reduced. Furthermore, the Chamorro government was strongly influenced by the Catholic Church and implemented a number of policies to return women to a more traditional societal status. Women activists took advantage of the Sandinista loss of power to organize themselves autonomous from any political party or existing organization. The new women's organizations were democratic with a horizontal structure. Women, this chapter will argue, had to fight harder to regain what they lost once the FSLN was voted out of power. The combination of economic liberalization and the introduction of policies influenced by strong Catholic values eliminated many of the women's options. The increase in employment and absolute and relative poverty combined with a loss of reproductive freedoms created extensive hardships for the vast majority of women in Nicaragua. In response to these new needs, several women's organizations built parallel institutions to address women's concerns, specifically in the area of women's health.

The women's movement and the FSLN, 1979–90

While the FSLN was in power, AMNLAE was part of its governmental structure. Consequently, AMNLAE had access to state-supplied resources and government contacts, assisting it in fulfilling many of the Sandinista promises of social justice and gender equality. However, AMNLAE's goals were vague and set by the FSLN rather than by women, and the FSLN did not always make gender equality a primary concern. Women, frustrated with AMNLAE, began to form their own organizations, critical of the Sandinista gender policies. These organizations were nonetheless Sandinista and followed the general Sandinista ideology. These new groups broadened the scope of the debate about women's issues while still allowing cooperation with the FSLN. Thus, women's needs and concerns were publicly debated although they were not addressed fast enough or democratically enough for the activists. Despite clashes between the women's movement and the FSLN, women's options and opportunities slowly improved throughout the 1980s.

To understand why the Sandinistas responded to some of the women's demands but not others, it is necessary to analyze how the mobilization of women benefited the government. The FSLN had two interests that influenced its relationship with women and the women's movement. The first was its social justice policies of improved education, health care, and economic sovereignty. Women benefited greatly from the programs, campaigns, and opportunities that these policies offered. Second, the FSLN explicitly targeted women in order to consolidate itself as a legitimate government.[1] These two interests were mutually reinforcing, i.e. what improved women's standard of living tended to increase their support for the new government. It was clear, however, that women's demands which did not fulfill these two interests were given little government support.

The FSLN took power from the dictatorship of the Somoza family which had held it since 1936. Anastasio Somoza Garcia ruled either overtly or through his puppets until 1956 when his son, Luis Somoza Debayle, assumed power. In 1967 Somoza Garcia's second son, Anastasio Somoza Debayle, became president and kept the position until 1979. The Somozas were able to maintain themselves in power by making alliances with the traditional ruling elites, by sharing money through patronage appointments and access to public funds, and by co-opting the labor movement before repressing it. Despite the sharing, Somoza became the largest landowner in Nicaragua.

By the time that Somoza fled the country and the FSLN declared victory, the anti-Somoza coalition was a complex web of alliances and groups representing all ranges of the political spectrum. Truly, the revolution was popular in both meanings of the word. The unity of the alliance, however, proved fragile in the task of rebuilding a bankrupt and wartorn nation. Somoza had absconded with all the money in the treasury, leaving the incoming government $1.6 billion in debt to various banks, international funding institutions, and countries. As a result of the fighting, damage to property alone came to US$470 million. There were approximately 600 000 people left homeless or with badly damaged homes, and per-capita productivity fell to 1962 levels. Due to the insurrection and related epidemics, 40 000 to 50 000 people died between 1977 and 1979.[2]

Immediately after gaining power the FSLN began implementing its ideology of social justice which improved women's options. In 1979 the FSLN banned sexist advertising and began a literacy crusade to educate the populace. A year later the FSLN began a health campaign. In 1984 legislation was passed that legislated equality in the family. In 1985 women's rights were codified into the new Nicaraguan constitution. The FSLN passed laws so that women could own land, belong to agricultural cooperatives, and receive equal pay for equal work.[3] The FSLN also set up women's cooperatives and trained women in business management.[4] The profile of women in power rose markedly as FSLN appointed women to positions of authority. By the mid-1980s women held 21 percent of the positions in the Ministry of the Interior and 15 percent of ambassadorial and international positions.[5] The Ministers of Police and Health were also women, and for five years the Speaker of the National Assembly was a woman. These government initiatives directly benefited women. Illiteracy dropped from 33 percent to 12 percent. As a result of improved health care, infant mortality declined from 120 to 80 per 1000 livebirths, malaria cases dropped by 50 percent, polio was eradicated, and the percentage of the population with access to health care rose from 28 percent to 80 percent.[6]

Economically, women moved into higher-paying jobs although the sexual division of labor still existed.[7] To increase women's productivity the FSLN established women's cooperatives. The fact that they did not flourish can be blamed on the war, the economic reality of running cooperatives, and Sandinista economic policies.[8] Politically, however, these numerous reforms and initiatives strengthened women's support for the FSLN. Women joined Sandinista organizations, especially AMNLAE, because it was involved in implementing many of these public policies. AMNLAE's members were the backbone of the 1979 literacy crusade and 1980 health campaign.[9] In 1985 AMNLAE organized 600 grassroots meetings, attended by over 400 000 women, to listen to women's concerns and to include their demands in the constitutional negotiations.[10] Thus, AMNLAE mobilized women across Nicaragua to help rectify social inequality. It was possible for AMNLAE to become highly organized so soon after the revolution due to its pyramid structure. The building blocks of AMNLAE's organization were groups, or chapters, of 30–100 women from every town. The cities had more than one chapter. Each chapter was made up of work committees of several women from every factory, farm, school, state institution, market, hospital, and neighborhood. Each chapter elected a representative to attend municipal and regional meetings. These representatives elected an executive committee, the members of which in turn reported to national headquarters.[11] From 20 activists in Managua and Matagalpa at the time of the revolution,[12] AMNLAE grew in little over two years to 25 000 members in 802 chapters throughout the country.[13] This structure was effective in mobilizing women, disseminating information and achieving short-term focused goals.

Despite its success, by 1981 AMNLAE recognized that its structure was alienating grassroots participants who felt that their voices were not being heard.[14] It also recognized that it lacked its own goals outside the needs of the revolution.[15] However, AMNLAE was incapable of altering its situation because it was controlled by the FSLN.

The FSLN was a party, mass movement, government, and state.[16] AMNLAE was part of its corporatist structure. As such, AMNLAE's structure, language, and goals had to fit with the needs of the whole. Thus, its leaders were delegated, its concerns were revolutionary, and women's needs were placed within a wider social context.

Occasionally AMNLAE did resist the directives of the FSLN. AMNLAE used its extensive membership to organize grassroots meetings in support of controversial legal and constitutional reforms that favored women.[17] Due to women's overwhelming public attention on these issues, the FSLN reluctantly passed legal reforms and improved women's rights in the new Nicaraguan constitution.

Any leverage that AMNLAE had over the FSLN was eroded by the 1984 election. The Sandinista Party won with a 67.9 percent majority, with women constituting 62 percent of the electorate.[18] Of the representatives elected, 19.6 percent were women. Only one of the 13 elected women was not from the FSLN.

The FSLN had not only consolidated its support among women, it had also consolidated its legitimacy as a democratic government. Thus, one of the Sandinista's interests in responding to women's needs had been fulfilled. Meanwhile, the high cost of fighting the civil war postponed the possibility for continued investment in social justice issues. By the mid-1980s the government's interest in targeting women's issues and organizing women declined sharply. For example, a 1985 study estimated that only 10 percent of the urban need for daycare was met and there were no immediate plans to change the situation.[19]

Sandinista women began forming their own women's groups to lobby the government and the rest of society for change. These new groups kept women's issues alive, informed and persuaded the public of alternative political approaches, and encouraged AMNLAE and the FSLN to change. In 1988 the Women's Secretariat of the Confederation of the Nicaraguan Professionals Heroes and Martyrs (CONAPRO H-M) began. Composed of middle-class, educated, Sandinista women, CONAPRO H-M sponsored two national conferences, in 1988 and in 1989, to publicize existing women's legal rights and to demand more. Both conferences received extensive media coverage and CONAPRO H-M's proposals were debated in the National Assembly.[20] In 1988 the first women's health clinic, *Ixchen*, opened in Managua. Led by another high-ranking Sandinista woman, it offered an alternative approach to women's health care. Women were encouraged to take ownership of their own bodies by participating in the decision-making, and by being informed about health issues, including reproductive health.

Because their leaders were widely known and their actions debated, *Ixchen* and CONAPRO H-M broke AMNLAE's control over the women's movement, facilitating the rise of women's collectives and instigating a broader discussion of gender politics. Both CONAPRO H-M and *Ixchen* were more horizontally organized than AMNLAE, and CONAPRO H-M was fully democratic. AMNLAE's leadership, trying to regain its lost monopoly of the women's movement, petitioned the FSLN to democratize AMNLAE. They hoped that an elected leader could better represent women's interests to both the FSLN and Nicaraguan women. The

FSLN promised such an election in 1989. It reneged on this promise, citing the need to focus all efforts on the 1990 general election. But the FSLN promised to democratize AMNLAE after the election.[21] In 1991 the FSLN again reneged, appointing another general secretary. This time, however, the FSLN was no longer in power, having lost the 1990 election. Without a guaranteed access to those in power, and without funding from the new government, this final breach of promise gave the discontented rank-and-file the opportunity to split from AMNLAE and create a women's movement autonomous from partisan politics.[22]

By the time of the 1990 general election women had become sufficiently equal in Nicaraguan society that the opposition coalition, the National United Opposition (UNO), had a woman as its leader. This equality was almost entirely due to the economic, political and social policies enacted by the Sandinista government.[23] Women gained positions of power and prominence in all aspects of Nicaraguan public sphere. In the private sphere women acquired legal equality.

Structural adjustment, conservative Catholicism, and women, 1990–96

With the election in 1990 of Violeta Chamorro and her coalition government, UNO, the economic structure of Nicaragua changed. While the goal of the government may have been to create a liberal democracy based on competitive individualism and free-market mechanism,[24] the reality of making that transition was daunting. Her coalition was unable to maintain its unity, forcing Chamorro and her Prime Minister, Antonio Lacayo, to create an alliance with the single largest party elected to the National Assembly, the FSLN. In 1991 the government publicly negotiated a *concertacion* with the FSLN party, Sandinista unions, and business groups. In exchange for protection of the Sandinista agrarian reform and some worker-owned businesses, the Sandinistas agreed to keep civil unrest to a minimum while the economy was undergoing its structural adjustment. This alliance required a renegotiation a year later, only to have the alliance prove politically irrelevant after the right usurped the National Assembly of 1993, and after constitutional talks jammed the National Assembly for another legislative session. The alliance crumbled in 1995 as the FSLN distanced itself from the Chamorro government in preparation for the 1996 election.

The Chamorro government privatized the economy through a harsh neoliberal program. This rapidly eroded the gains that women had achieved under the FSLN. Their standard of living fell, their employment opportunities diminished, and their access to government-funded programs and services was reduced. The increase in women's economic hardship made it problematic for women to organize due to lack of free time and difficulties exiting the home.

Just as the FSLN implemented its social justice beliefs through public policies, so too did Chamorro use public policy to implement her religious beliefs. Her conservative Catholicism, with its emphasis on men caring for women in the home, legitimated the withdrawal of services essential for women's health and well-being. In addition, the triumph of conservative Catholicism, with its female ideal of motherhood, undermined the movement for women's rights.

When Chamorro took power the economic situation was dire.[25] Inflation was 13 490 percent, the debt stood at $10 billion, four times the gross domestic product (GDP), and underemployment and unemployment was 45 percent.[26] The Chamorro government implemented a plan to stabilize the Nicaraguan economy by privatizing industries, cutting government expenditures, and imposing monetary stability. The stabilization plan promised to bring hyper-inflation to zero within the first 100 days of office by devaluing the Nicaraguan currency 116 percent and then fixing it at parity with the American dollar. After a year, inflation dropped to 865 percent, and the currency was devalued a further 400 percent. In January 1993 it was devalued a further 20 percent and transferred to a crawling peg. Despite devaluations, by 1994 the currency was overvalued by at least 50 percent, hampering exports. Nonetheless, inflation dropped to 12.4 percent by 1993.

The SAP also tightened the supply of money, reducing credit from the Central Bank to the commercial banks by 36 percent in 1991 and a further 50 percent in 1993.[27] There is the suggestion that this adversely affected small and medium-sized producers more than larger ones.[28] The macroeconomic structure was further stabilized by massive foreign aid.[29] Between 1990 and 1994 Nicaragua received more aid per capita than any other developing nation.[30] The US alone transferred US$541 million in the first two years of the government.[31]

The debt, however, continued to grow at a rate of US$520 million per annum, and by 1994 was eight times the GDP.[32] This is despite the fact that the government significantly reduced its expenditures. Subsidized goods such as food and electricity had their prices raised to market value. The size of the military was reduced. The public workforce was reduced.[33] State funded medical care was reduced with the best hospitals privatized. Prescription medicine was no longer free.

Funding for education was reduced at all levels of the school system, and children who attended public schools had to pay tuition even at the primary level.

None of these measures served to reactivate the economy. The government hoped that liberalizing the economy to allow free market mechanisms to dominate would promote business. Thus, the SAP privatized state companies, which accounted for 30 percent of the country's GDP;[34] deregulated the economy; ceded control of foreign trade and banking to the private sector; and dropped tariffs from 43 percent to 15 percent.[35] The government encouraged wages to be cut to promote exports and, in 1994, reformed the Labor Code to "flexibilize" labor, i.e. erode labor rights.[36]

The economy did not show any growth until 1994. Between 1990 and 1994 per-capita GDP contracted 12.4 percent with a cumulative growth rate over the same period of 1.9 percent.[37] The increase in exports in 1994 came primarily from an unanticipated frost in Brazil which improved the terms of trade for Nicaraguan coffee.[38] Exports increased again in 1995, raising per-capita GDP for the first since the revolution. However, underemployment and unemployment still pushed 70 percent.[39] If the economy does continue to grow, as some independent economists have predicted, then the benefits will likely not be distributed throughout the population. With the economic restructuring:

> the Chamorro administration's naive and uncritical faith in market forces and its general disregard for market imperfections injected a bias which – by default, if not by design – favored large businesses and strengthened the positions of monopolies, oligopolies, and the business elite of the government.[40]

In particular the SAP eroded women's standard of living. "Typically with orthodox adjustment, the distributive impact fell heavily on women."[41] Of female-headed households 72.5 percent were below the poverty line compared to 67.5 percent of male-headed households; 81.3 percent of female-headed households reduced food consumption compared to 75 percent of male-headed households.[42]

Unemployment and underemployment rates for women were higher than for men[43] because women dominated public sector employment.[44] Up to 70 percent of government layoffs were women.[45] In addition, the recession also forced women from private sector employment because they were the first to be laid off.[46]

Nonetheless, women had to work to feed their families. It is estimated that 65 percent of urban families and 82.6 percent of rural households were headed by women.[47] Women responded to the economic crisis by actively searching for alternative means of earning money. In comparison, men adopted a posture of resignation to or avoidance of the family crisis.[48] Thus, even with a male head of the household, there was no guarantee the family would be supported.

Not merely were women the hardest hit by the economic downturn, but their standard of living was further reduced by the cutbacks in social infrastructure spending. Social services that predominantly affected women were reduced or withdrawn. Government-sponsored training initiatives and community programs, such as daycare, were eliminated. The lack of training left women unable to alter their skill level, trapping them in poverty. The lack of community programs increased women's child-care concerns and reduced their ability to exit the home in search of paid employment.

In addition to the economic restrictions that women had to overcome, women also had to overcome the return of the established Catholic Church to a position of influence within the government. Chamorro was a devout Catholic. The connection between the President and Nicaragua's Catholic Cardinal was so close that "she [was] perceived by some as a surrogate for the Cardinal Miguel Obando y Bravo."[49] In addition, many of the President's top advisors belonged to the neo-conservative sect, the Charismatics.[50] The Church and the government cooperated to present to the Nicaraguan people an idealized form of femininity consisting of a mother happy in her home. Chamorro spoke frequently of the need to return women to the home, of strengthening the family, and of reestablishing traditional values.[51] Through public policy, Chamorro's government restricted women's access to reproductive control. Fertility, maternal mortality, and illegal abortions increased, compounding the difficulties women faced in escaping poverty.

In 1992 the government reformed the Penal Code through Law 150. With this law procreation became the basis for all sexual relationships. When the constitutionality of this law was contested before the Supreme Court, the justices decided that the entire purpose of the sex act was to procreate and the purpose of procreation was to increase the population and thereby to build a state. Therefore, anything that interfered with procreation was a crime against the state.[52]

Included in this bill was a legal punishment for procuring or performing abortions. This was the first time that abortion had been criminalized in Nicaragua. Doctors and midwives were both explicitly denied the legal right to provide an abortion.[53] After the penal code was passed, the government moved against abortion clinics that insisted on staying open, and jailed a doctor and a woman for performing and receiving, respectively, an abortion.[54]

These legal and public policy changes had the support of the traditional Catholic Church in a complementary campaign to ban birth control and abortion. The traditional Catholic Church in Nicaragua, politically active since the mid-1970s, became increasingly concerned with sexuality.[55] Bolstered by its close connection to the government and the Pope's designation of 1994 as the "Year of the Family," the Catholic Church in an "unprecedented move"[56] organized a march for life on the National Assembly to strengthen laws against abortion. The Church organized the march as part of a Father's Day celebration, the significance of which was not lost on the population: if abortion and contraceptives were legal then men could not be fathers. If men were not fathers then men could not be men.[57] The Church "intensified its campaign against abortion and 'unacceptable' campaigns which encouraged the use of contraceptives."[58]

The result of this policy was an increase in maternal mortality.[59] The United Nations Fund on Population estimated that for every 100 000 livebirths there were 300 maternal deaths.[60] However, that official number is dubious. Statistics on maternal mortality are suspect because of the lack of reporting. It has been estimated that only 47 percent of pregnant women were registered by the health-care system,[61] meaning that less than half the births and deaths were known to the system. However, it was reported that, every 36 hours, one woman died as a result of being pregnant.[62] Women died from lack of proper health care before, during, or after a pregnancy. Women had to pay for the use of gloves, medicine, instruments, food, and sheets, yet many lacked the money to do so. Thus, there was a high degree of septic infections.[63]

In addition to poor maternal health care, women also died from an increase in the number of abortions. This also stemmed from economic liberalization and conservative Catholicism. Seventy-five percent of women who aborted said they did so because they lacked the money to raise children.[64] Women, however, eschewed therapeutic abortions because of the Chamorro government's strong religious pronouncements and the enforcement of anti-abortion laws; women's requests for therapeutic abortions dropped to 3 percent of that in 1989.[65] Instead, women aborted outside the health-care system, risking their own health. One-third of all maternal deaths, it was estimated, were from self-induced abortions, although again it was difficult to confirm this because the fear and shame of discovery ran high.[66]

The implementation of Catholic values as public policy changed the nature of state intervention into the private lives of its female citizens. While the Sandinista government intervened for its own purposes, its intervention consisted of reforming the social structure to increase the economic and political options available to women. In fact, the Sandinistas offered women so much freedom that they chafed at the restrictions imposed on their own organization. In comparison, state intervention under Chamorro was subtler. The state did not directly or indirectly manipulate civil society organizations, leaving an impression of a civil society autonomous and free from state machinations. Yet the options for women narrowed. The state placed a primacy on women's procreative role within the home, and then reinforced this role through the educational system, penal reform, Supreme Court directives, health-care cutbacks, and religious mobilization. Women's ability to freely choose any number of paths was reduced to one as the state shepherded women into the private sphere of childrearing.

Ironically, women were simultaneously propelled from the home and into the marketplace in order to feed their family. Economic liberalization reduced the standard of living and economic options of most Nicaraguans but it especially affected women. Neo-liberalism also reduced access to the mechanisms through which poor women improved their standard of living: education, training, and social services, such as day-care and health care. Thus, women's opportunities to alter their economic position were also reduced. This reduction of women's economic position meant that women had insufficient money to afford private maternal health care. The consequences were deadly for women, especially given that the government limited women's control over their own reproduction. In this context of state-led economic liberalization and conservative Catholicism, women's options and rights, both economic and personal, were significantly constrained. Indeed, the government's statements and public policies threatened the very physical survival of women.

The women's movement in Nicaragua, 1990–96

Initially, AMNLAE was not helpful in resisting the policies of the new Chamorro government. After the 1990 general election AMNLAE and the Sandinista women's unions declared themselves autonomous from the FSLN party. However, their new mandates were unclear.

Consequently they had limited ability to respond to the rapid changes imposed by the Chamorro government. By 1994 AMNLAE had reforged its mandate and began a rotating microbank to assist women's economic development in the informal market.

In the interim, without leadership from well-known women and access to the party in power, the women's movement reinvented itself from the ground up. The women's movement democratically chose, in two national conventions, the exact opposite structure from AMNLAE. Familiar with the weaknesses inherent in a centralized and controlled mass movement, the autonomous women's movement risked a radically horizontal organization. It based itself on women's cooperatives, neighborhood centers, and health and law clinics. Decentralizing the movement encouraged local control, which boosted women's participation. By 1995 there were approximately 150 grassroots organizations.[67] These disparate groups were united through networks focused on a specific issue. The networks generally stretched across the nation, informing interested individuals and groups, and growing slowly. Frequently AMNLAE, other Sandinista women's unions, and women's think-tanks also joined. Rarely, however, did the Chamorro government's women's institute (INIM) join.

Networks allowed for maximum decentralization, participation and democratization of the women's movement while nonetheless promoting unified action on common problems.[68]

Networks offered multiple leadership positions as each group brought its own leaders, identities, and strategies to the problem. As a Nicaraguan women's activist explained, networks were a way

> to coordinate, act together, discuss and define strategies and positions to take with respect to the state and other women. At the same time, we all address [women's] problems in an autonomous way through our own institutions or organizations.[69]

In order to achieve their goal, networks generally created a group within themselves specifically to organize around the goal. This group generally drew upon women from a number of collectives to do much of the organizing, fund-raising, writing, media communication, etc. While this could be seen as a hierarchy, it might more accurately be understood as a division of labor as each woman's loyalty would be to her own collective.

Once established, a network would hold a large and well-advertised public conference where its goals and strategies would be discussed and decided upon. Conferences also served to inform the general public of women's needs. During these conferences the network organized future events, petitions, and other public activities to either lobby the government or influence civil society. Examples of influential networks included the Women Against Violence Network, the Self-Convocated Women's Network, and the Women's Health Network.

The Women's Health Network became powerful for a number of reasons. The most important reason was that the rate of maternal mortality was so high that it mobilized and united women who otherwise might not have been politically active. Regardless, without its unique history and structure, the Women's Health Network would never have attained its position of influence.

The complex structure of the network was rooted in the Sandinista years. Health-care clinics were initially promulgated and organized by AMNLAE and then by Sandinista women's legal and health groups. In 1987 AMNLAE began building women's centers throughout Nicaragua, two dozen of which offered health care. *Ixchen,* a women's center devoted to health care and law, opened its doors in 1988. This led to the creation of smaller clinics. This was especially so after the 1990 election because the Chamorro government refused to hire health care workers trained by the Sandinista government. These workers, already trained and experienced in mobilizing the grassroots for community health care and integrating health care into neighborhood organizations, then established neighborhood health-care clinics.

By 1992 there were 16 organizations bringing services to 52 centers in all the departments (provinces) including the autonomous states of RAAS and RAAN.[70] Of these, 61.6 percent offered reproductive health care; 29.9 percent offered education (training through theater and workshops); 4.7 percent offered psychological support; 3.9 percent offered legal advice, mostly dealing with domestic violence.[71]

Because the network was based on local collectives or clinics and staffed by volunteers, its financial needs were surprisingly modest. Foreign aid was a steady if not a generous supply. Foreign aid did not appear to overtly affect the goals or organization of the women's movement. When women were asked if the funding source affected what they said or did, they replied no.[72] Foreign aid would have difficulty directing the operations of the Women's Health Network because its grassroots groups were too small and disparate, the network too amorphous, and the variables too numerous for any one agency to control. Thus, the Women's Health Network may not have been able to attract large sums, but by its very organizational structure it was able to avoid being subordinated to the wishes of the funding agency.

The final and most critical reason why the Women's Health Network became so influential was that the network operated within the women's movement and included neighborhood movements or autonomous groups from all other movements. It did not operate within the government's health-care structure. There was no pretense that health was anything but a woman's issue. For example, in 1991, *¡Sí Mujer!*, a women's clinic, opened its doors with a statement of its intent:

> Our center, integrated into the feminist movement, is demonstrating the solidarity power of women together in the fight to the end for the defense of all our rights that permit us to advance along the path to the conquest of our primary vindication of gender.[73]

It was so well accepted that women's health was the cornerstone of the women's movement that the leaders of the health clinics in the 1990s were as well known and as widely quoted as AMNLAE's leaders were in the 1980s.

The clinics integrated health care with feminism through a conscious effort to cultivate an egalitarian structure (doctor–nurse and practitioner–patient). The women attending the clinic were encouraged to become informed and active about their health, intending to "transform the passive object to an active subject with her own identity."[74] The transformation from passive object to active subject was a critical precondition for women's participation in the Women's Health Network.[75] This approach was at odds with the government's health-care structure which actively discouraged workers from broadening their treatment to include health-care education. For example, health-care workers who volunteered to educate women about sexually transmitted diseases were threatened with dismissal.[76]

Ironically, the government's attempts to roll back women's health care and control women's reproduction focused the women's movement around this issue. In the first six months of the Women's Health Network there were 216 activities.[77] This may seem like a great many, but given the number of collectives in the Women's Health Network it becomes more understandable. The Women's Health Network included neighborhood clinics from all the major urban centers. It also included AMNLAE's women's centers and independent women's centers. In addition, there were other networks that intersected with the Women's

Health Network: the Professional Women for Development Network, the NGOs which worked in health, and the think-tanks. The Women's Health Network intersected through its member NGOs, the government's health organizations and the Ministry of Health, which kept the government informed about the network's activities, the number of collectives involved, and the potential participation upon which the network could call.

The Women's Health Network integrated all the collectives and clinics yet allowed each unit to mobilize, organize, and publicise around the issue of health as each saw fit. For example, the health clinic *¡Sí Mujer!* operated autonomously from the network, collecting its own information, producing its own pamphlets, and mobilizing its own constituents. The Women's Health Network "María Cavallieri"[78] also operated autonomously. In comparison, five of the Managua neighborhood women's centers banded together to organize joint marches and utilized the same pamphlets. Each group highlighted issues that it thought were the most relevant.

Despite the differences, all the groups came together in conferences that made explicitly political demands. In order to do so effectively, the Women's Health Network split into two sub-networks, one explicitly directed at influencing the formal political sphere and one at mobilizing civil society. Although the latter was superficially less political, the Women's Health Network tried to influence societal norms through raising public awareness about women's health. Thus they tried to inform, attract, and mobilize a constituency in support of policy changes. The former sub-network persuaded through information; the latter through symbols. The symbol that became the focus of the fight between the second sub-network and the government was Mother's Day.

Mother's Day became a battle over a symbol of motherhood. Mothers were traditionally glorified as an ever-sacrificing, passive nurturer, and that symbol was one of the most powerful symbols in Nicaragua. Mother's Day, 28 May, is a celebration marked by an excess of sentiment, even by Nicaraguan standards. The sub-network wanted to change the image to one of an unfair and occasionally deadly battle against an inadequate and oppressive health-care system. Thus, the sub-network named itself and renamed Mother's Day to Maternal Mortality Day. The means through which Maternal Mortality Day hoped to influence the general public was media information, eye-catching demonstrations, and public petitions presented to the President of the National Assembly.

Beginning on Mother's Day 1992, mothers whose daughters had died in childbirth carried signs with their daughters' names and their ages at death. The women paraded up and down in front of the National Assembly and marched up the street. It was a stunning display of pain-initiating action. Many of the female deputies, including Azucena Ferrey (conservative and head of the Women and Family coalition), came out to talk to the women. The event was covered by the local media, granting the demonstration legitimacy and helping to transform Mother's Day into a political issue.

The Women's Health Network was able to change the symbol of motherhood because it pointed out its inconsistencies: the paradox of worshiping motherhood in a society that then denies maternal health care. The Women's Health Network also offered an alternative: changed laws and public policies, abolished patriarchy, and new democratic practices. These alternatives were presented by the second subnetwork of the Women's Health Network

The second sub-network focused on international discussions and agreements, and on issues such as population and development. This group used the international women's movement, international agreements, United Nations (UN) Conferences, NGOs, and pressure from supportive nations, such as the USA, to lever the government–church alliance. This sub-network positioned itself as the nexus between the international community and domestic political organizations.

The connection between the lack of development, mortality, and birth rate has been clear for a number of years, especially since the concern was widely articulated at UN conferences,[79] specifically the September 1994 UN Cairo Conference on Population and Development, the March 1995 Social Summit in Copenhagen and the September 1995 UN Beijing Conference on Women. Preparation for all three conferences had been a permanent action for the Women's Health Network since 1993. The sub-network organizing for the international conferences had the overarching plan of changing the Nicaraguan government's stance on reproductive health. For example, in December 1993 there was a preparatory conference in Managua for the Cairo and Beijing Conferences. This was repeated in April 1994 where resolutions were reached. Ostensibly for the UN conferences, the resolutions in fact exhorted the Nicaraguan government to create new public spaces, consult with the women's movement, broaden grassroots participation in preparatory meetings for the upcoming UN conferences, and communicate with the preparatory groups organizing for the UN conferences.[80]

This sub-network had limited success. Due to the split at the Cairo Conference where the Vatican and seven other nations refused to sign the convention, international pressure on the Chamorro government was diluted. As a result the Chamorro government also refused to sign resolutions in support of women's reproductive rights. The Beijing Conference was more successful due to international consensus on key issues. The Chamorro government signed the final agreement. However, this did not change any domestic policies.

The Women's Health Network clearly demonstrates the difference between the Sandinista women's movement in the 1980s and the women's movement under a state-imposed neo-liberalism and Catholicism. The FSLN enacted a health-care program. The FSLN chose the health-care goals and then acted its program through government-trained health-care specialists. To deal specifically with women's health, the government-initiated women's centers operated by the Sandinista women's group, AMNLAE. This mass organization assisted with health campaigns. Women's health care was not a political issue *per se* but rather one of the FSLN's social justice issues. With health care the Sandinista women's movement and the FSLN shared the same general goal and thus were able to act cooperatively and effectively to improve women's health.

In comparison, women's health care became the paramount political issue under neo-liberal economic policies and conservative Catholicism, and an example of the hostile relationship between the women's movement and the Chamorro government. The Women's Health Network, which emerged to address this issue, was an example of the successful reorganization of the women's movement to confront the Chamorro government's policies. The movement included neighborhood groups, newly autonomous women's centers, newly created health and law clinics, and women's research groups all united into networks. Because women were forced back into the home with fewer possibilities of exiting it, and fewer resources to expend on organizing, the women's movement had to decentralize, dealing with local needs at the local level. In effect the women's movement went to where the women were. Networks were horizontally organized with each group granted a high degree of autonomy from the others. Leadership was democratically chosen, as was the extent and form of involvement.

The Women's Health Network was uniquely situated to become the most effective and visible arm of the new women's movement. The clear and evident need for maternal health services galvanized the women's movement into organizing health clinics. In addition, the hostility of the Chamorro government to Sandinista-trained health-care workers drove experienced personnel into grassroots mobilizing. Finally, there was considerable international interest in women's health which generally worked in favor of the network.

The Women's Health Network, as a result, was able to shift public discussion of maternal health from ignorance to awareness, changing societal values in the process. The Women's Health Network engaged the government about its policies and their effects on women, publicized the issue of the government's reproductive health policies, and forced the government to acknowledge its demands in public forums both domestically and abroad. The neighborhood clinics that were part of the Women's Health Network informed women about their bodies, replaced some essential services, and empowered women to be active participants in their health care. The accessibility and success of the Women's Health Network increased women's participation in it, strengthening the Nicaraguan women's movement. Unfortunately, the network could not replicate the level of health that women enjoyed under the Sandinistas. For that matter, the women's movement could not replicate the social and political options available to women under the Sandinistas. The women's movement in the 1990s attempted valiantly, but with limited success, to ameliorate the worst of the impact of the government's economic and social policies, and regain what women had lost under Chamorro's government.

In October 1996 the Constitutionalist Liberal Party under the direction of Arnoldo Aleman won the federal election in Nicaragua. The Liberal Party took 44 seats with 46 percent of the popular vote while the FSLN remained the second-largest party with 37 seats and 36.5 percent of the popular vote. Aleman is continuing with neo-liberal economic policies and at the end of 1997 announced the signing of the second Enhanced Structural Adjustment Facility (ESAF) with the IMF.

Aleman has also created a new organization, the National Institute for Women and Children, which has already gained a reputation for its arch-conservativeness. Although Aleman himself has not evoked Catholic sentiments or worked with the Church as Chamorro did, the Church remains an influential political actor in contemporary Nicaragua. Given the effect of the continued neo-liberalism combined with traditional Catholic values, the social and economic status of women will remain profoundly unequal to men with a pronounced impact on women's health and well-being. It is likely that the women's movement will continue to respond to women's needs as best it can.

Conclusion

State-led economic liberalism and conservative Catholicism fundamentally altered the Nicaraguan women's movement. Neo-liberalism reduced or eliminated the mechanisms through which women had pursued social justice: social services and economic infrastructure. State-led Catholicism further reduced or eliminated the mechanisms through which women pursued equality and freedom: law and reproductive control. Neo-liberalism may have freed up market mechanisms but conservative Catholicism changed the nature of government intervention into women's private lives by increasing regulation over women's reproduction. The combination of the government's two initiatives, one economic and the other ideological, reinforced the negative consequences each had for women. Women's economic and physical survival was threatened while their capacity to participate in political affairs to change that situation was reduced.

Women's concerns shifted from social justice, equality and freedom to survival. The women's movement refocused their goals and strategies. Fortunately, the democratization of the women's movement was already well under way by the time Chamorro took power. The women's movement democratized itself in response to Sandinista patriarchy, rather than in response to the Chamorro government's neo-liberal economic policies. The fact that the women's movement democratized under economic liberalism was a mere coincidence. Regardless of the economic policies that the Chamorro government pursued, the women's movement would have democratized once the Sandinistas no longer could use its resources and state power to control the movement.

Women's increased and democratic participation in an autonomous civil society organization could be taken for a strengthening of civil society. The essence of the optimism about a strengthened civil society lies in its potential, first to lobby the government for change, and secondly to improve the lives of its participants. In the case of the Nicaraguan women's movement, much of the academic optimism about civil society under neo-liberalism is misplaced. In the first instance, the women's movement, a powerful civil society organization, was not powerful enough to overturn the policies affecting women. Fiscal restraint continued to undermine women's economic survival, and government control over women's reproduction continued to threaten their physical survival. In the second instance the Nicaraguan women's movement could not replace all government-supplied services –

even after restructuring itself when it was forced to focus on finding foreign funding and replacing key services. Ultimately, what a civil society organization can expect to achieve in an era of state-led neo-liberal economic policies and conservative Catholicism is a limited replacement of some government services, but not the democratic capacity to influence the state's public policies.

Notes

1 Maxine Molyneux, "Mobilization without Emancipation? Women's Interests, the State, and Revolution," in Richard Fagan, Carmen Diana Deere and José Luis Coraggio (eds), *Transition and Development: Problems of Third World Socialism* (Berkeley, CA: Monthly Review Press and Center for Study of the Americas, 1986), p. 228.
2 John A. Booth, *The End and the Beginning: The Nicaraguan Revolution* (Boulder, CO: Westview Press, 1985), p. 183.
3 Clara Murguialday, *Nicaragua, revolución y feminismo (1977–89)* (Madrid: Editorial Revolución SAL, 1990), p. 126; David Close, *Nicaragua: Politics, Economics, and Society* (London and New York: Pinter, 1988), pp. 160–1.
4 Ibid., p. 105.
5 Michelle A. Saint-Germain,"Women in Power in Nicaragua: Myth and Reality," in Michael A. Genovese (ed.), *Women as National Leaders* (Newbury Park, CA: Sage, 1993), pp. 70–102, 125.
6 Helen Collinson *et al.* (eds), *Women and Revolution in Nicaragua* (London: Zed Books, 1990), pp. 98–9; 122.
7 Gary Ruchwarger, *Struggling for Survival: Workers, Women, and Class on a Nicaraguan State Farm* (Boulder, CO: Westview Press, 1989).
8 Geste Dijstra, *Industrialization in Sandinista Nicaragua: Policy and Practice in a Mixed Economy* (Boulder, CO: Westview Press, 1992); Richard Stalher-Sholk, "Stabilization, Destabilization, and the Popular Classes in Nicaragua, 1979–1988," *Latin American Research Review*, **25**, 3 (1990), 55–88; Linda Mayoux, "From Idealism to Realism: Women, Feminism, and Empowerment in Nicaraguan Tailoring Co-operatives," *Development and Change*, **23**, 2 (April 1992), 91–114.
9 AMNLAE 1983.
10 Collinson *et al.*, op. cit., p. 143.
11 Association of Nicaraguan Women Luisa Amanda Espinoza, *Una mujer donde este debe hacer revolucion*, translated by Latin American Working Group (Toronto: Latin American Working Group, 1983), p. 3.
12 Murguialday, op. cit., 102.
13 Susan E. Ramirez-Horton, "The Role of Women in the Nicaraguan Revolution," in Thomas W. Walker (ed.), *Nicaragua in Revolution* (New York: Praeger, 1982), pp. 147–59, 154–5.
14 Women's International Resource Exchange (WIRE), *Nicaraguan Women: Unlearning the Alphabet of Submission* (New York: Women's International Resource Exchange, 1985), p. 24.

15 Beth Woroniuk, "Women's Oppression and Revolution: the Nicaraguan Debate." Unpublished paper for Canadian University Students Overseas, 1987, pp. 5, 8.
16 Harry E. Vanden and Gary Prevost, *Democracy and Socialism in Sandinista Nicaragua* (Boulder, CO: Lynne Rienner, 1992).
17 Gary Ruchwarger, *People in Power: Forging a Grassroots Democracy in Nicaragua* (South Hadley, MA: Bergin & Garvin, 1987), p. 190; Doug Brown, "Sandinismo and the Problem of Democratic Hegemony," *Latin American Perspectives*, **17**, 2 (Spring 1990), 39–61 at p. 53; AMNLAE, 1983, p. 11.
18 According to most independent observers the election was clean and fair given the constraints of the counterrevolutionary war being waged against the FSLN government.
19 Woroniuk, op. cit., p. 7.
20 Sarah Morgan, "Toward a New Revolution in Health: the Emergence of Alternative Health Care Networks in Nicaragua since the 1990 Elections." Unpublished undergraduate thesis, Hampshire College, Amherst, MA, 1995, p. 80.
21 Margaret Randall, *Sandino's Daughters Revisited: Feminism in Nicaragua* (New Brunswick, NJ: Rutgers University, 1994), p. 211.
22 *La Gente*, #57, weekly cultural supplement to *La Barricada*, 15 February 1991, p. 11.
23 Saint-Germain, op. cit., pp. 80–2.
24 William I. Robinson, *Promoting Polyarchy: Globalization, US Intervention and Hegemony* (Cambridge: Cambridge University Press, 1996), p. 251.
25 Mario Arana, "General Economic Policy," in Thomas W. Walker (ed.), *Nicaragua without Illusions: Regime Transition and Structural Adjustment in the 1990s* (Wilmington, DE: SR Books, 1997), p. 85.
26 José Antonio Ocampo, "Collapse and (Incomplete) Stabilization of the Nicaraguan Economy," in Rodiger Dornbusch and Sebastian Edwards (eds), *The Macroeconomics of Populism in Latin America* (Chicago, IL, and London: University of Chicago Press, 1991), p. 331; Arana, op. cit., p. 86.
27 Dijstra, *Technology Questioned: Ideology and Interests in Nicaraguan Economic Politicies* (The Hague: Institute for Social Sciences Working Papers, no. 213, 1996), p. 193; *Envio*, Managua, Universidad CentroAmerica, **12**, 150 (January 1994), 7.
28 Ibid., p. 193; Richard Stahler-Sholk, "Structural Adjustment and Resistance: the Political Economy of Nicaragua under Chamorro," in G. Prevost and H. E. Vanden (eds), *The Undermining of the Sandinista Revolution* (New York: St Martin's Press, 1997), p. 91; Florence Babb, "After the Revolution: Neoliberal Policy and Gender in Nicaragua," *Latin American Perspectives*, **23**, 1 (Winter 1996), 43.
29 Nestor Avendaño, *La Economia de nicaragua: El Ano 2000 y las Posibilidades de Crecimiento* (Managua: Nitlipan and CRIES, 1994), p. 9.
30 *Envi*, **14**, 166 (May 1995), 5.
31 Robinson, op. cit., p. 242.
32 *Central America Report*, Guatemala City, 15 December 1994, 4.
33 Rose Spalding, *Capitalists and Revolution in Nicaragua: Opposition and Accommodation, 1979–1993* (Chapel Hill, NC: University of North Carolina, 1994), pp. 166–9.

34 Arana, op. cit., 86.
35 Dijstra, *Technology Questioned*, p. 17; Stahler-Sholk, "Structural Adjustment and Resistance," p. 90.
36 Ibid., p. 92.
37 *Envio*, **13**, 151–2 (February–March 1994), 6.
38 Stahler-Sholk, "Structural Adjustment and Resistance," p. 94.
39 Arana, op. cit., p. 92.
40 Ibid., p. 84.
41 Stahler-Sholk, "Structural Adjustment and Resistance," p. 96.
42 Anna Maria Fernandez Poncela, "The Disruptions and Adjustment: Women in Nicaragua," *Latin American Perspectives*, **23**, 1 (Winter 1996), 56.
43 Stahler-Sholk, "Structural Adjustment and Resistance," p. 96; Cynthia Chavez Metoyer, "Efficiency or Burden Shifting? Gendered Outcomes of Stabilization and Adjustment Policies in Nicaragua." Paper presented at Latin American Studies Association International Conference, Washington, DC, 28–30 September 1995, p. 19.
44 Collinson *et al.*, op. cit., p. 32.
45 Nan Weigersma, "State Policy and the Restructuring of the Women's Industries in Nicaragua," in Nahid Aslanbeigui, Steven Pressman and Gale Summerfield (eds), *Women in the Age of Economic Transformation: Gender Impact of Reforms in Post-Socialist and Developing Countries* (London and New York: Routledge, 1994), p. 201.
46 Fernandez, op. cit., p. 52.
47 Ibid., p. 55.
48 Ibid., p. 56.
49 Saint Germain, op. cit., p. 91.
50 Penny Lernoux, *People of God: The Struggle for World Catholicism* (New York: Viking, 1989), pp. 400–1.
51 Saint Germain, op. cit., p. 97.
52 Supreme Court Decision #18, 7 March 1994, II:iii.
53 *Barricada International* (English), **359**, 3 (March 1993), 19.
54 Sonia Agurto Vílchez, *Analisis de la Situation Actual de la Población Nicaragüense: Un Analisis de Género* (Managua: Autoridad Sueca para el Desarrollo Internacional, 1992), p. 14.
55 Lernoux, op. cit., pp. 368–9.
56 *Central America Report*, 1 July 1994, p. 4.
57 Maria Angelica Fauné, "Mujeres y Familias Centramericanas Principales Transformaciones y Problemas: Desde la Pespectiva de las Mujeres." Paper presented at a conference conducted by Swedish Aid, San José, Costa Rica, May 1994.
58 *Central America Report*, 1 July 1994, p. 4.
59 Interview with Patricia Ríos, nurse with Sí Mujer. Interview with author, 25 April 1994.
60 Fondo de Poblacin de las Naciones Unidas (FPNU), *America Latina y el Caribe: Carpete de Informacion* (1993), p. 15.
61 Ana María Pizarro, "Encuentro Nacional de Mujeres Sobre Politicas de Población y Desrrollo: Por un Futuro Digno y Seguro para Todas y Todos" (Managua: Fondo de Población de las Naciones Unidas, 1993), 3.
62 *La Barricada* (Spanish), 27 May 1994, p. 4.

63 Interview with Ríos.
64 *La Gente*, #227, 29 July 1994, p. 4.
65 Morgan, op. cit., p. 70.
66 Lois Wessel, "Reproductive Rights in Nicaragua: From the Sandinistas to the Government of Violeta Chamorro," *Feminist Studies*, **17**, 3 (Fall 1991), 541–2.
67 Myra Pasos, "The Women's Movement in Nicaragua Today." Speech, York University, Toronto, Canada, 6 March 1995.
68 Vargas, op. cit., p. 3.
69 *Barricada International*, **362**, 6 June 1993, 23.
70 María Dolores N. Ocón, "Los Servicios Alternativos de la Salud de la Mujer en Nicaragua." Paper prepared for the Pan-American Health Organization, Women, Health and Development Program, Managua, September 1992, p. 3.
71 Ibid., p. 7.
72 Interviews with Ríos; Alicia Seranno, Director of Projects, AMNLAE, interview with author, 21 April 1994; Marta Munguia, Director of Accion Ya!, interview with author, 28 March 1994; Maria Teresea Blandon, Co-ordinator National Feminist Committee (CNF), interview with author, 6 January 1994; Rogalia Ramos Sevilla, Director, *8 marzo*, interviews with author, 20 and 28 June 1994.
73 *¡Sí Mujer!* (1994), pamphlet. Managua, op. cit., p. 1.
74 Ocón, op. cit., p. 5.
75 One of the components of this transformation is to encourage women to be community leaders (interview with Sevilla).
76 Field notes from organizational meetings with Xochiquetzal, Managua, 1994.
77 *La Gente*, #121, 12 July 1992, 11. An activity is anything that is formally organized for the public from a meeting to a demonstration.
78 This started up in the Matagalpa region before the Managua-based network began. Thus, it is its own mini-network and refuses to be wholly absorbed into the larger network.
79 The connection between birth rate, development and infant mortality was known to *Voces de Mujeres*. In its petition circulated among participants at the December 1993 Preparatory Conference for the Cairo Conference on Population and Development, the connection is clearly made in the first paragraph of the Preamble (*Voces de Mujeres*, Declaricion de las Mujeres Sobre Las Politicas de Poblacion. Petition for presentation at the Cairo Conference on Population and Development 1993, 1).
80 *La Barricada*, 14 May 1994, p. 8.

Conclusions: Politics of Democratization – the Force of Civil Society

Remonda Bensabat Kleinberg and Janine A. Clark

The adverse effects of economic structural adjustment policies (SAPs) on the socioeconomic landscape of the developing world, while varied, are challenging the political assumptions upon which SAPs have been designed and implemented. While the World Bank generally abstains from political statements and interference in the domestic politics of its members, it has recognized the need (if only to ensure the success of SAPs) to understand the political context and effects of Bank lending and programs. Economic programs are premised on assumptions concerning their short- and long-term political impact. In the 1990s the dominant view of policy-makers and advisors in the international financial institutions (IFIs) is that SAPs are compatible with the introduction of political liberties and democracy in the developing world. Within this understanding of economic and political change, civil society is seen to play a crucial political role. Simplistically stated, as the state withdraws from previous spheres of activity, civil society is expected to organize and take over "abandoned" state activities, and press for greater political liberalization and decentralization of decision-making powers.

Yet the political realities of economic reform are proving to be antithetic (if not contradictory) to the political expectations of domestic and international policy-makers. The successful implementation of SAPs as economic policy in the world's poorer countries has almost required their application at any political cost, particularly if contingent international loans are forthcoming. As the contributors to this volume suggest, governments have been amenable to the utilization of anti-democratic methods in their implementation of SAPs. In this context civil society has been unable to fulfill its theoretical role, and, as the case studies of Uganda, Tunisia, Mexico, Nicaragua, and

Argentina show, the wholehearted execution of structural adjustment and economic liberalization has shielded these countries from international pressure to *fully* democratize.

Economic reform and the rise of civil society

Yet the impact of economic reform on civil society's role in the processes of political liberalization and democratization in the regions of Asia, Africa, the Middle East and Latin America has not been exclusively negative. Despite what the contributors to this volume found to be an inherent tension between economic liberalization and democratization – one that expresses itself in increased state authoritarianism and often a lack of tolerance and democracy at the grassroots level – we do find evidence pointing to the fact that civil society does play an important role in popular participation and democratization. The current popularity and usage of the term "civil society" in relation to political liberalization and democratization are based on more than wishful thinking. Rather, we find "success stories," albeit limited, as civil associations take advantage of political configurations and directly and indirectly force political change upon their governments and states.

To a certain extent economic reform programs as dictated by international financial institutions in the developing world can be accredited for fostering an environment conducive for increasing numbers of NGOs and other civil associations. In 1991 Kenya's principal donors, the World Bank, the US, Britain and France, for example, demanded that substantive political reforms be implemented before the country's loans were released. At the grassroots level, Afolayan, Clarke, Clark, Dicklitch, and Jones document a significant rise in NGOs as direct result of structural adjustment programs in Nigeria, the Philippines, Egypt, Uganda, and Indonesia, respectively. Between 1984 and 1993 NGOs grew by 148 percent in the Philippines. According to Clarke this increased the number of NGOs to 58 000, the most vocal and active of which were the human rights groups. In Indonesia the number of NGOs registered with the Ministry of Home Affairs reached 3251 by 1989 – double the number registered in 1985. As Jones states, the period of fantastic NGO growth was "the same period [marking] the beginning of Indonesia's long, slow 'structural adjustment' process."

Jones, furthermore, notes that financial deregulation in Indonesia has enabled NGOs to access credit to foster income-generating projects and local savings schemes. This was also the case in the Philippines where, by 1989, the Task Force Detainees of the Philippines (TFDP), the largest human rights NGO in the developing world, administered more than 100 income-generating projects. It also ran its own rehabilitation center for released prisoners, while other grassroots organizations provided food and medicine to detained political prisoners.

In Egypt the World Bank has created the Social Fund for Development as a means of alleviating some of the negative socioeconomic effects of structural adjustment. The Social Fund has dispensed large sums of aid to NGOs; a few have been flooded with money.

The private business sector, emerging as the most powerful segment of Mexico's civil society, certainly has benefited from economic reform. Economic deregulation in Mexico has allowed business to "bypass" the state in accessing credit and seeking out international finances as well as joint ventures and other international strategies. As a result of the "liberalized" environment Mexico can boast an increase in its own multinational companies operating in the United States and Europe. Indeed, large farmers in Tunisia have ready access to resources in order to modernize their farms from subsistence to agricultural exports. Economies of scale, and mechanization, are fast replacing traditional farming methods. In these senses economic reform programs have been a boon.

More commonly, however, IFIs can be attributed with a largely negative form of inducement or incentive in the creation of civil associations. The resultant economic austerity and harshness of economic reform has spurred the creation of civil associations out of necessity and the rising and painful disparity between rich and poor. Self-help associations have become especially important for the survival of the developing world's poor. In Egypt government officials admit the crucial role local NGOs play in providing social-welfare services to Egypt's poor. In many districts throughout the country, volunteer NGOs are the primary caregivers and driving force in local development. In Tunisia there is an impressive array of associations providing services to the poor: 509 charitable, relief and social associations and 126 associations for development, alongside thousands of other cultural, artistic, friendship, and scientific organizations. In another example the hostility of the post-Sandinista government in Nicaragua to Sandinista-trained health-care workers fueled grassroots mobilizing.

In the Philippines human rights organizations now target "developmental aggression" – the human rights abuses associated with macroeconomic reform – as the focus of their efforts. These groups have shifted from a definition of human rights that rests on state transgressions in terms of civil liberties, including torture and violence, to one that is based on economic interventions by the state or private entities with the support of the state that endanger the interests of communities economically or otherwise. In a similar vein, Loh documents that in Malaysia the Federation of Malaysian Consumers Associations launched a campaign against the increase in sewage rates upon privatization of the national sewerage system.

Analogically, the arrival of foreign firms as an integral component of economic reform has triggered the formation of NGOs. In Indonesia workers organized strikes in reaction to the discrepancy between Indonesian workers and foreign managers. Workers fought back when the government attempted to regularize a minimum-wage law. NGO activists were pivotal in the creation of local unions. Similarly, Dicklitch notes that railway and medical workers in Uganda respectively organized strikes in response to suspended or low wages.

Civil associations: goals and achievements

Other than increasing their sheer numbers, civil associations can boast more qualitative achievements within the context of economic reform. In the Philippines, for example, human rights groups were instrumental in pressuring the government to create the Presidential Human Rights Committee in 1988. This resulted in speedier investigations, direct hearings with human rights victims, and more vigorous prosecution of violators. In another example, in 1992, Philippine human rights NGOs were largely responsible for the repeal of the Republic Act 1700, an anti-subversion act, and the creation of the government's Fact-Finding Committee on Involuntary Missing Persons.

In Malaysia, NGOs forced the government to withdraw an amendment to the Societies Act which required NGOs to declare themselves as "friendly" or "political." NGOs furthermore pushed the hand of the government to create the Royal Commission of Inquiry in the early 1980s in order to investigate the illegal expropriation of public funds. Malaysian NGOs also achieved a series of successes in the field of environmental protection. In 1986 and 1987, respectively, NGOs launched major campaigns to prevent amendments to the Official Secrets Act and to the Constitution.

In Jordan, Moore notes that economic reform has brought with it the creation of new channels between business groups and the government for lobbying and consulting. Business lobbying was able to effectively amend an unpopular sales tax law. Similarly in Mexico, the drive for a North American Free Trade Agreement between the United States, Mexico and Canada, allowed business associations more routine, conventional access to government officials and the negotiating table. Economic liberalization helped to establish the basis for a more transparent, formal, consultative mechanism between government and the business class.

Examining the other half of the business equation, Patroni elaborates on the strides labor has made in Argentina. In 1988, labor successfully renegotiated with the government the new law of professional associations more in line with terms amenable to the labor leadership.

In Egypt NGOs are largely responsible for the drafting of a new Law 32 – the main government instrument of control over NGOs; it will soon be presented in parliament. Similarly, in Nicaragua, health-care workers and health-care activists were gratified to witness the Chamorro government, under NGO pressure, sign the 1995 Beijing Conference on Women agreement. This, however, did not result in any real policy changes at home.

Most importantly, NGOs have played a vital role in public education and consciousness-raising. In her discussion of Nicaragua, Isbester succinctly exemplifies the type of educational role played by NGOs throughout each of the case studies. She states:

> The Women's Health Network ... was able to shift public discussion of maternal health from ignorance to awareness, changing societal values in the process. The Women's Health Network engaged the government about its policies and their effects on women, publicized the issue of the government's reproductive health policies, and forced the government to acknowledge its demands in public forums both domestically and abroad. The neighborhood clinics that were part of the Women's Health Network informed women about their bodies, replaced some essential services, and empowered women to be active participants in their health care.

The politics of state–civil society relations

Nevertheless, not all of the initiatives launched by civil associations, or even their achievements, have been meaningful or long-lasting. Operation Lallang in Malaysia, for example, brought NGO momentum to an abrupt halt in 1987. Operation Lallang introduced draconian security laws that allowed for the detention of persons without trial. In Tunisia the regime utilizes strategies to prevent associational autonomy and successfully maintains social, financial and political control over its dense network of civil associations. In Jordan, newly created business–state channels for policy participation have been weak and largely ineffective, and in Mexico only big business has made great strides while micro, small, and medium businesses are struggling to survive: they have little access to government credit or resources.

However superficial governments' responses to civil society pressures may be, they do indicate the recognition of governments' need, at a very minimum, to pay lipservice to the legitimate rights of their citizenries. Ultimately, governments' actions are bound at the two extremes by the need to avoid (for fear of) destabilizing civil unrest. At the one end, governments, fearful of rising popular activism and the potential loss of power, reverse the political liberalization and democratization that gave rise to increasing political opposition through legislation or state violence. At the other end, civil society's negative reaction to state abuses and increasingly authoritarian measures incurs a limited implementation of civil liberties and democratic rights.

In Malaysia, for example, Operation Lallang was introduced by a government nervous about its ability to remain in power. Similarly, in order to control social and political instability, state agents in Tebourba, Tunisia, have promoted neo-traditional social organizations, and reinforced attachments to traditional patronage and kinship networks and to Islamic values and institutions. As a result of this strategy, primordial identification has increased while associations representing cross-cutting cleavages have declined during economic reform. Contrarily, in Egypt, state excesses in the attempt to eliminate Islamic radicalism and violence have galvanized pro-democracy groups, as well as supporters of the government party, to force the government to introduce greater political liberties and democracy. This has brought strange bed-fellows together as secular liberals and moderate Islamic groups join forces to pressure the government.

Ultimately, the unique, domestic political configurations created by the exigencies of economic reform, political stability, and popular demands in each country produce often unpredictable avenues or opportunities for civil activism and influence – ways of eluding state control. For instance, the Kenyan church has taken advantage of government reluctance to deny the right of religious expression and has played a paramount role in the struggle for pluralism and government accountability. In Nigeria, as Afolayan asserts, popular sentiment against the state was reflected in cartoons and folklore. Cartoons especially became an effectual means of expressing anti-government opinions. Before long they began to appear in books, pamphlets, and leaflets, and were distributed throughout the country in the various dialects and indigenous languages. Story-telling and song were also found to be a way to express popular sentiment outside the watchful eyes of the government-controlled electronic media.

In Egypt many social-welfare NGOs have avoided the restrictive arm of the Ministry of Social Affairs (MOSA) by declaring themselves as religious or Islamic and therefore registering with the Ministry of *Awqaf* (a religious endowment ministry) instead. Other NGOs use informal, personal patron–client relations to circumvent MOSA regulations and restrictions. More importantly, some NGOs have taken advantage of new laws introduced with the IMF structural adjustment program. These NGOs register themselves as civil or professional companies and not as non-profit NGOs. While they must pay taxes as companies, they avoid MOSA. In an example from Jordan, upon occasion business has been able to take advantage of divisions within the government. Business was able to significantly amend a 1993 tax law due both to the support of opposition parties in parliament and the differences of opinion between relevant ministries over the tax law.

While these political dynamics create political opportunities that are not generalizable, the contributions in this volume do point to certain common factors that provide crucial support enabling civil associations to achieve their goals. While these factors cannot be said to be either necessary or sufficient, they do contribute to our understanding of the conditions under which civil society may promote political liberalization and democratization under economic reform.

Civil society and international linkages

Certainly, in a majority of cases, international political support and/or international donor aid have provided civil associations with the inspiration, means and exposure to continue their work and be heard. The philosophical shift in the 1980s by international donors toward funding NGOs has had a significant impact in the number of viable NGOs in the developing world. As Isbester states in the case of Nicaragua, international development agencies have increasingly directed funds at women's grassroots health-care initiatives because the key to economic, community and sustainable development lies with women.

Clark points out how foreign donations provide a significant proportion of Egyptian NGOs' budgets. Indonesian human rights groups were inspired by the Polish Solidarity movement and formed an organization with the Indonesian equivalent of the name in 1990. In the early 1990s the Indonesian workers' groups received prominent support from Dutch and American trade unions. Similarly, in Uganda, local NGOs joined the international NGO, the Structural Adjustment Participatory Review Network, in its efforts to ameliorate the negative effects of economic reform. Moreover, Ugandan NGOs, the number of which increased exponentially since 1986 and include the Uganda Community Based Child Care Association (UCOBAC), the Youth Alliance for Development Co-operation (YADCO), and several others, have received funding by the Danish Development Agency (DANIDA), the United States Agency for International Development (USAID), and the African-American Institute (AAI).

In the Philippines in 1996 the denial of an entry visa to Nobel Peace Prize winner José Ramos Horta, in order that he may attend an NGO-organized human rights summit, attracted international attention and concern. In Malaysia the opposition party, the Parti Bersatu Sabah, was partially able to achieve victory at the polls in 1986 due to international bad press which the state government party and its illegal tactics (including bombings) received. Embarrassed by the press, the federal party withdrew funding and support for the state leadership.

The World Council of Churches has provided both educational and financial support to the Kenyan clergy in their efforts to educate the citizenry on voting rights and other political matters. Police or military crackdowns on peaceful meetings have led to international condemnation and, at least on one occasion, the suspension of international loans to the state. International support has thus, in several cases, given NGOs a "face" in the international community and has enabled them to avoid being completely shut down via legislation. International financing has also aided NGOs in continuing their work.

This is not to suggest that foreign aid is unproblematic. As Isbester discusses, too often foreign donors have been accused of attempting to steer the direction of the NGO, creating significant tension between their goals and those of the indigenous organizations. The case studies in this volume reinforce, however, that, under the right circumstances, foreign aid can provide invaluable support to civil associations. International linkages and international aid can have positive results, if only because they offer financial resources and international clout not otherwise available to the groups and associations.

The Women's Health Network in Nicaragua, for example, has successfully avoided the pitfalls of tied-aid. Isbester states that

> [f]oreign aid did not appear to overtly affect the goals or organization of the women's movement. When women were asked if the funding source affected what they said or did, they replied no. Foreign aid would have difficulty directing the operations of the Women's Health Network because its grassroots groups were too small and disparate, the network too amorphous, and the variables too numerous for any one agency to control. Thus, the Women's Health Network may not have been able to attract large sums, but by its very organizational structure it was able to avoid being subordinated to the wishes of the funding agency.

Clearly, as will be discussed in greater depth below, civil associations' organizational structure is partially defining the strength and autonomy from foreign donors of civil associations.

Civil society and the media

The success of certain initiatives by civil associations can also be partially attributed to their effective use (and the support) of the media, including conferences. The Women's Health Network in Nicaragua regularly holds large and well-advertised conferences in order to discuss its goals and strategies, and to inform the general public of women's needs. As Isbester states, during these conferences the women organize future events, petitions and other public activities to either lobby the government or influence civil society. The network uses pamphlets, photocopies and magazines to spread its message.

In Nigeria, private newspapers and weekly magazines doubled between 1985 and 1993 after the Babangida administration repealed the anti-press decree of the former military regime. While the government managed to amass press support with its rhetoric on human rights and democracy, it was not prepared for widespread media criticism once the administration tightened its hold on power. Afolayan contends that the media "directed all its venomous criticism [at Babangida], using terms and phrases that made press watchers even in relatively freer societies like the United States shudder." That the government clamped down hard on the press is not surprising. However, in order to elude government officials, some magazines altered their format to tabloid and eventually went underground. The underground tabloid *Tempo* became the "most popular and most widely read tabloid in the country."

The Philippine human rights NGO, TFDP, is heavily supported and consulted by journalists, politicians, and diplomatic officials. It has a long history of credibly researching and documenting human rights abuses in the country. In Indonesia, despite the geographic challenges of large numbers of widely dispersed islands, NGOs have been able to forge nationwide coordination and action through the use of improved communications technology. In the 1990s the All-Indonesia Workers Union has managed to get the vast majority of incidents of government harassment into the domestic and international press.

In Jordan the Amman Chamber of Commerce elections have become highly publicized. The public and government view these elections as bellwethers of municipal and parliamentary elections. In their opposition to the peace process with Israel, leaders of the business associations heavily used Jordan's daily newspapers in expressing their views.

Civil society and the politics of networking

Comparisons across the developing world also indicate that the organizational structure of civil associations and their networks can strongly influence their effectiveness. As mentioned above, Isbester largely attributes the success of the Women's Health Network in Nicaragua to its decentralized, radically horizontal structure that allows for maximum participation and democratization of the movement, while simultaneously promoting unified action to common problems.

In Uganda 14 diverse, indigenous NGOs are horizontally linked under the umbrella organization, the National Organization for Civic Education and Election Monitoring (NOCEM), formed in 1993. While limited in their political activities (narrowly defined), their organizational make-up has enabled the training of thousands of civic educators and the monitoring of elections in 20 pilot districts in Uganda.

In Indonesia the broad-based alliance of workers, organized by the Center for Indonesian Workers' Struggle (PPBI) could well have brought down Prime Minister Soeharto in the final analysis. The consistent mobilization and networking of workers in domestic and international firms, and its growing popularity, led to the formation of the PPBI's own political party, and the latter's increasing association with other NGOs and student groups. Labor rights became associated more and more with democratization. While increasing arrests in 1994 and 1995, and the crackdown on labor demonstrations in 1996, weakened NGOs' ability to organize, the role of NGOs and student activism has yet to be fleshed out in the fall of the Soeharto leadership.

Isbester and Clarke credit the diverse variety of NGO goals – targeting local and specific concerns as well as broader national ones – as important to their success. According to Isbester, the grassroots approach of the Nicaraguan Women's Health Network is the most critical reason why the network became so influential and came to be the cornerstone of the women's movement:

> the network operated within the women's movement and included neighborhood movements or autonomous groups from all other movements. It did not operate within the government's health-care structure. There was no pretense that health was anything but a woman's issue.

In the Philippines the TFDP targeted a wide variety of issues in the 1980s. It helped establish the Philippine Alliance of Human Rights Advocates, and it supported local grassroots organizations such as Families of Victims of Involuntary Disappearances, Mothers and Relatives against Tyranny and Repression, the Association of Ex-Detainees against Detention and for Amnesty. It also supported a broad range of socioeconomic and political issues, such as foreign debt, overseas workers and agrarian reform, and formed nationwide networks.

Horizontal linkages between multi-issue, non-hierarchical autonomous associations have had tremendous success in accomplishing their goals and setting agendas. However, as stated above, there is no one recipe for success. Differing domestic political configurations present different obstacles and opportunities for civil associations. In terms of the former, the Philippines' human rights movement has been subject to government co-optation of certain associations within the movement. On the other hand, in the case of Kenya, activists from churches across the country have organized into a strong hierarchical association. The Church has provided a certain moral protection from the government and, as a result, activists have been able to continue their work.

Civil society and historical political experiences

Finally, several of the authors indicate that the prior political experience of the civil associations (or lack of it) influenced their effectiveness. The political experience of the Women's Health Network under the Sandinista government, for example, created an invaluable foundation for the network in terms of consciousness-raising and the acquisition of organizational and other skills.

Comparably, Patroni traces the historical "framework" within which Argentina's trade unions continue to operate. The relationship between labor and the government in Argentina is strongly delimited by the political experience of key players beginning with the creation of the labor movement prior to the rise of Peronism. The trade unions exhibit a long history of seeking the support of the state in the resolution of labor conflicts and to enter politics if necessary – even if the "trade-off" is one of its own autonomy.

Political experience of both business and labor has come in the form of historic corporatist arrangements in Mexico. Indeed, corporatism has continued to define state–private sector–labor relations, albeit without the entitlements. Powerful business interests have found these "traditional" channels exceptionally convenient for expressing their interests. So, while the private sector has sought a liberalized economic environment, the call for the dismantling of semi-autonomous, legislated business and workers' associations is not forthcoming. For labor, access to state support remains crucial, and for business, access to policy-making and labor leadership is paramount.

Political experience can work to propel the interests and goals of a movement, association, or sector, as may be the case, forward more efficiently. Members take political skills with them to whatever association they may join. In other examples the foundation, pattern or channels for negotiation between civil society and the state are established historically over time. While economic restructuring may affect the dominant players in civil society, or the form these players take, the patterns of negotiation largely remain. As the cases of Mexico and Argentina exhibit, this may work to the advantage or detriment of civil associations.

Yet historical political experiences may also play a more overtly negative role. Without making a cultural statement the history of political violence in many countries in the developing world has had an impact – often detrimental – on today's civil societies. Democratic political skills may be inhibited rather than developed. In terms of the case studies in this book this is most evident in the chapters on Africa.

In Kenya the history of physical abuse by the state has engendered a "'high-strung' passionate culture of opposition" that has fostered violence between opposition groups. Consequently, ethnicity has become a mobilizing force used for narrow political ends. Combined with the inadequacy of non-violent channels, there is no cohesive support base for democracy to exist, as ideology has not become the binding force of political parties. The ethnic/class differentiation has been the driving force behind economic liberalization.

In Uganda the legacy of ethnicity has been made clear in Dicklitch's succinct article. Uganda, as Dicklitch contends, "is infamous for the bloodshed that characterized much of its post-colonial history." Ethnicity, fueled by British colonial preferential treatment to its largest tribe, entered the realm of Ugandan politics, making patronage, coercion and repression the primary route to political stability. A "unifying nationalism" that gave legitimacy to the state was absent, and violent confrontation became the instrument by which ethnic groups would attempt to secure control of the political system.

The context of violence in these African countries has partially tended to inhibit the progress of civil associations as they seek a democratic voice; but as the Nigeria case shows, while violence, ethnicity, and regionalism have not escaped the political landscape, pro-democracy sentiment is being cultivated through popular culture and the underground media. Though still in its infancy, civil society is re-emerging and finding sophisticated approaches for expressing its political views and democratic demands.

Cross-sectoral generalizations

Twelve case studies of the developing world scarcely constitute a sample upon which to make cross-sectoral generalizations. However, we do see a trend in this volume's case studies between and across sectors as a result of structural adjustment and economic liberalization. Of all sectors studied, business, or the landowning class as is the case in Tunisia, is benefiting the most under economic reform. In Mexico and Malaysia business is gaining substantially more access to policy-making and, overall, has stronger ties to the government than ever before. In Jordan, while a pact between the state and business has yet to fully emerge, business is becoming more vocal in national debates over economic policy. As the case of Mexico demonstrates, deregulation has made it easier for the business class to access international and domestic partnerships, as well as credit and other financial and capital resources.

As the case of Tunisia suggests, large farming interests are closely tied to political officials – if they themselves are not officials. Indeed, at the local level political control is orchestrated through an alliance between large landowners and officials in the Ministry of the Interior. Resources, credit, below-market input prices and technological support is easily available to the landowning class, and the latter have been the only real beneficiaries of structural adjustment in Tunisia.

Labor and the peasant class, on the other hand, have been steadily losing ground *vis-à-vis* historical ties to the state. In Argentina, as well as Mexico and Indonesia, labor is losing its entitlements as well as access to state support – a powerful defeat for historically influential syndicates and unions. Labor is, moreover, weakening through loss of jobs, wages, man-hours and the recent reconfiguration and/or dismantling of strong industry-wide unions. Small farmers and peasants in Tunisia have little access to credit, have faced increasingly declining prices for livestock and crops and have steadily lost access to state lands as a result of privatization. Moreover, politically connected big farmers have kept the farmer's union from meeting or voicing opinion over agricultural policy, leaving small farmers no avenue for support.

As the consequences of structural adjustment are being *felt*, an interesting shift seems to be occurring within the human rights NGOs. Human rights groups in the Philippines, Uganda, and to some extent, Indonesia, for diverse reasons, have been circumscribed to the *non*-political realms. In the case of the Philippines where political detainees have been released, and state violence against the opposition has decreased considerably under the Ramos regime, human rights are now being defined to include economic and social rights. Poverty, as a result of "development aggression" and the preponderance of the capital class, has become the target to human rights activities. With the creation of official human rights boards to deal with human rights abuses as traditionally defined, human rights NGOs have become increasingly marginalized.

The demands of foreign capital in Indonesia's low-wage, deregulated market have steadily diminished labor's share of the economic pie. In response, human rights groups have increasingly added labor rights to their agenda and have demonstrated openly alongside workers against companies such as Nike. Clearly, here, human rights have taken on more economic and social meaning as well.

In Uganda, on the other hand, human rights groups are under a constant threat of being dismantled by the state if they enter into politics. Instead, under the NOCEM umbrella, these groups are focusing on educational roles and election monitoring. In all cases, rather than dismantling, we see a re-definition of the roles and functions taken on by human rights organizations.

Finally, the case studies in this book that deal with social-welfare associations, including those run by and directed to women, commonly share increasing responsibilities without adequate government support, planning or a policy framework within which to direct or coordinate their activities. Government responses to social-welfare NGOs range from abandonment to neglect to hostility. NGOs have been significantly overwhelmed in their ability to help the poor, due both to the increasing numbers of people seeking their help, and the inadequate human and financial resources of NGOs'. Legislation has further hindered NGOs capacity for doing so by limiting resources and controlling their activities. Even IFIs have understood the impact structural adjustment has had on the poor sectors of society, particularly since one of the primary adjustments is the slashing of health care. In the case of Egypt the World Bank has tried to alleviate NGOs' burden through funding assistance.

Perhaps more than any other type of civil association, social-welfare NGOs have also mutually displayed ingenuity in evading the restrictive arm of the state in order to better address the needs of the poor. Thus, we see in both the Kenyan and the Egyptian case studies how educational and social-welfare groups respectively maintained a non-political status and refused to register as NGOs by taking advantage of the privileged status of religious institutions. In Kenya groups registered themselves as church organizations as opposed to NGOs in order to elude legislative restrictions to their activities. In Egypt social welfare groups have similarly registered as non-political entities, and prefer to register as religious, civil, or professional companies in order to secure their autonomy, and avoid MOSA and Law 32. The mosque, like the church in Kenya, has become an important source of aid as well as a haven for circumventing state control and diminishing state resources.

For diverging reasons, women's groups have also had a great deal of success "surviving" in Nicaragua. The importance of "democratizing" the women's movement from within, and the emphasis on networking, has led to a not-easily diminished, broad-based movement in Nicaragua. Though not effectual in gearing policy to their interests, the women's movement there has been able to develop other strategies at the grassroots level to help address their immediate needs, and, perhaps, survive on a much longer-term basis.

Democracy and social justice

The case studies in this volume strive for different goals and face dissimilar obstacles. While Kenya, Nigeria, and Uganda struggle to achieve the most basic political and democratic rights, the Philippines is a case of democratic consolidation. Yet the chapters share a common understanding of the need for a more comprehensive form of democracy, one that goes beyond an institutional definition of democracy to one that addresses social justice, if democracy is to take hold and survive in the developing world.

Certainly, in the case of Egypt, for example, structural reform is directly resulting in greater numbers of poor, the needs of whom NGOs simply cannot address adequately. While there are many causes of Islamic militancy, economic hardship as a result of SAPs is providing recruits for groups striving to bring down the government and state through violent means. Strong-arm measures by the so-called democratic state to eliminate radical groups are dangerously undermining democracy as a contending ideology in Egypt and peaceful democratic change as a viable political option.

After over 60 years of relative social stability and (at times) coercive solutions to civil unrest, the Mexican state is resorting more and more to outright violence to quell the emergence of an increasing number of "revolutionary" movements. Violent confrontations, growing poverty, and the marginalization of an increasing number of unemployed and underemployed continue to represent the "dark" side of a country pursuing economic reform and integration into the global economy. Within this context democratization *per se* is seriously impeded.

Human rights groups in the Philippines have been the most forceful in demanding that economic reform be implemented with a human face. Philippine human rights NGOs are challenging both the acts of human abuses and the definition in order to include those abuses created by economic reform programs. These NGOs are most actively conscious of the political economy of reforms and the resultant socioeconomic injustices the poor are suffering.

At the heart of civil society's debates over the impact of SAPs, however, lies not only a discussion over the short- or long-term economic viability and beneficiality of SAPs. Rather, at its essence, is a debate over the nature of state–civil society relations and the degree to which economic reform programs are helping or hindering the democratization of those relations. Economic reform is taking place amidst little or no discussion concerning the future role of the state and/or civil society. Civil society is playing little, if any, role in the planning and implementation of SAPs, and despite the increasing responsibilities of civil associations, they have little knowledge of state plans for the future. In economically weak and politically unstable countries this neglect by governments and IFIs alike is problematic and precarious.

The long-term effects of economic reform on the democratization of state–civil society relations are yet to be revealed. Substantive democracy – one that provides access to the election booths as well as the institutions which will enable civil society to fulfill social and economic needs, is clearly absent in all the case studies presented here. Fundamentally, it is not enough to make changes solely to economic policy. It is the editors' opinion that, in order to democratize, *political* policies need to be legislated for that transition to occur, and states must address increasing poverty rates. The political and socioeconomic repercussions of not doing so are enormous.

Analyses on state–society relations remain state-centered and tell us little about the force and progress of civil society in its quest for democracy. Paradoxically, in order to gauge the qualitative changes made at state level, future research must continue to focus on changes occurring from "below." This is even more important if we are to understand precisely how economic reform empowers, enables, or weakens a society's ability to push for substantive political change in an inherently authoritarian or coercive climate.

Index

Printed in the United States
124011LV00001B/19-39/A

9 780312 230876